METHODS IN MOLECULAR BIOLOGY

Series Editor
John M. Walker
School of Life and Medical Sciences
University of Hertfordshire
Hatfield, Hertfordshire, UK

For further volumes
http://www.springer.com/series/7651

3D Sponge-Matrix Histoculture

Methods and Protocols

Edited by

Robert M. Hoffman

AntiCancer, Inc., San Diego, CA, USA
and
Department of Surgery, UCSD, San Diego, CA, USA

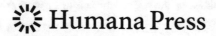

Editor
Robert M. Hoffman
AntiCancer, Inc.
San Diego, CA, USA

Department of Surgery
UCSD
San Diego, CA, USA

ISSN 1064-3745   ISSN 1940-6029   (electronic)
Methods in Molecular Biology
ISBN 978-1-4939-9272-0   ISBN 978-1-4939-7745-1   (eBook)
https://doi.org/10.1007/978-1-4939-7745-1

Frontispiece

Joseph Leighton, M.D., Chairman

In Memoriam:
Joseph Leighton
1921–1999

Father of 3-Dimensional Tissue Culture

This book is dedicated to the memory of A.R. Moossa, M.D., and Sun Lee, M.D.

Preface

The Dawn of Tissue Culture

Maintaining and growing cells and tissues outside of the body was a momentous achievement that began in the late nineteenth century [1]. Many scientists at that time did not believe that growing cells and tissues outside of the body would be possible.

Roux maintained the medullary plate of an embryonic chicken in a saline solution for several days, thereby establishing the principle that tissues could live outside body [1].

In 1907, Harrison [2] observed nerve fibers growing from the explants from the nerve fiber tips of the tadpole spinal cord on a lymph clot. Burrows used plasma clots instead to grow nerves and mesenchymal cells from chick embryos. Carrel and Burrows transplanted fragments of the original culture to clean slides with fresh plasma and the procedure could be repeated [3, 4]. This was perhaps the first "passage" of tissue cell culture.

Carrel [5] maintained fragments of connective tissue and beating heart in vitro for more than 2 months and transferred the cultures from medium to medium consisting of Ringer's solution and hypotonic plasma in order to maintain them long term. The culture was then sealed on a hollow slide, put in an incubator, and allowed to grow [1].

The Rous chicken sarcoma grew rapidly in culture [6]. The Rous sarcoma was grown in vitro and re-transplanted into a chicken where it formed a sarcoma. Spindle cells predominated in slow-growing cultures, and the round cells predominated in rapidly growing cultures. The round cells were perhaps the first observation of cancer stem cells [1].

The Dawn of Cell Culture

In 1923, Carrel [7] developed the first practical cell culture flask with a long sloping neck, which prevented contaminants from entering this flask. New medium could readily be added to the flask [1].

Tissue culture was revolutionized by the introduction of antibiotics in the 1940s [1].

Rous and Jones in 1916 [8] used trypsin to detach and obtain suspensions of cells from the cellular growth around explants, which were then passed in new culture vessels [1].

Earle established the continuously growing L-cell line [9]. George Gey established HeLa cells from a human cervical tumor [10].

Earle and Ham developed defined cell culture media [11]. Eagle developed a medium (Eagle's minimum essential medium [MEM]) with over 25 ingredients [12]. These media were supplemented with serum such as fetal bovine in order for cells to proliferate [1].

In 1948, Sanford, Earle, and Likely [13] cloned a mouse cell line, NCTC 929, in a capillary tube containing "conditioned" medium from mass cultures of growing cells [1].

Szilard said: "Since cells grow with high efficiency when they have many neighbors, you should not let the single cells know they are alone" which led to the "feeder cell" concept [1]. Puck [14, 15] X-irradiated the feeder cells that retained their metabolic activity for purposes of conditioning the medium, but the cells could no longer proliferate. Single cells were cloned on the monolayer of X-irradiated feeder cells. In 1956, a medium for growing HeLa cells was developed such that a feeder layer was no longer required for cloning [14, 15].

Science usually does not progress in a straight line and sometimes while seemingly going forward, goes backward. After the time of Carrel, monolayer cell culture became the dominant paradigm. Nutrient medium was being developed, and cancer cell lines were being established that grew very well as monolayers on glass and could be cloned as described above. These developments led to the birth of the whole new field of somatic cell genetics with pioneers such as Harris [16], Ephrussi [17], Puck [14, 15] and others.

Viruses such as polio could be grown in monolayer culture for vaccines and other purposes [18]. Molecular biology was born in the 1950s and although much of the early work was done in *E. coli*, it was natural for molecular biology workers to graduate to monolayer cultures of human and mouse cell lines.

The Dawn of 3-Dimensional Cell and Tissue Culture

Some workers, especially coming from a field such as pathology, understood that monolayer cultures had no connection to the structure of tissues and organs. Joseph Leighton, a pathologist, perfectly understood this concept [19–33] and thereby developed true 3-dimensional (3D) sponge-matrix histoculture, which could allow cells to self-organize to tissues and organoids to form in vitro.

However, 3D sponge-matrix histoculture never made it to the mainstream. It never became a paradigm despite its obvious superiority as a cell and tissue culture system compared to what has become, in the ensuing 70 years. A much simpler system became the paradigm of "3D culture," even though it is not really 3D (please see below).

What is termed "3D culture" is very popular now. Journals such as *J. Cell. Science* [34] and *Development* [35] have devoted whole issues to "3-dimensional culture" under the term "organoid" (please see Afterword, this volume). "Organoid" is an old term (meaning organ-like, using the Latin root "oides" which means similar). However, in the second decade of the twenty-first century, "organoid" has taken on a much more restricted meaning: "3D culture of organ-like structures initiated by stem cells" [36]. The leader of this new "organoid movement" is Johannes Carolus Clevers, from the Netherlands who goes by and published as "Hans Clevers." Hans is not only an outstanding scientist but also a very good strategist. He took advantage of the "avalanche" of research on embryonic stem (ES) cells [37], and induced pluripotent stem (iPS) cells [38], and Hans' very own stem cells (LgR5) [36], which he presents as an example of an "adult" stem cell (aSC), as opposed to the above-mentioned "embryonic" stem cells. Hans most astutely combined the stem-cell field with the "3D culture field" led by Mina Jihan Bissell, a Persian-American woman who obtained all her education at Harvard. Mina's father came from a long line of Ayatollahs. During Mina's postdoctoral years at Berkeley and the Lawrence Livermore Laboratory, in the late 1970s and early 1980s, she worked with two great scientists, Harry Rubin, who introduced her to cell culture, and Melvin Calvin, who provided her facilities and colleagues to help her use instrumentation which enabled precise incubation of cell cultures. At that time, the by-far dominant paradigm of cell culture was placing cells in plastic dishes or flasks where they grew as monolayers. Cancer researchers used monolayer culture to study cancer cell lines in order to compare them to normal cells, and a whole "industry" of such studies rose up in academic centers and pharmaceutical companies around the world.

In the middle 1980s, Hynda Kleinman, working at the U.S. National Cancer Institute, began using a laminar-rich extracellular matrix produced by Engelbreth-Holm-Swarm (EHS) tumor sarcoma cells [39] originally discovered by Orkin [40]. Hynda commercially developed this matrix, which came to be known as "Matrigel." Matrigel culture has become the very dominant paradigm for "3D culture" due in large part to Mina's use of Matrigel to culture breast cancer cells in "3D" [41, 42]. Using these cultures, Mina and her team claimed that upon treatment of breast cancer cells growing on Matrigel with antibodies to particular "integrins," the breast cancer cells became more "normal" [34]. Commenting on Mina's discovery, Jacks and Weinberg [43] and Abbott [44] (Figs. 1 and 2, respectively) proclaimed "the new dimension in biology" and Mina was anointed the founder and queen of "3D culture." For this work, Mina was elected to the National Academy of Sciences, the American Philosophical Society, and other honorary organizations and has received numerous awards and became the leader of the 3D culture field, claiming that 3D cell culture started with her research [34, 45].

Spending more than 50 years in science allows the observation of many fads that come and go and come back again. We are now back to before the 1950s with the current very popular 3D culture methods. At the present time, it seems that most scientists have either forgotten about or are unaware of sponge-gel matrix histoculture developed by Dr. Leighton.

News Feature

Nature **424**, 870-872 (21 August 2003) | doi:10.1038/424870a

Cell culture: Biology's new dimension

Alison Abbott[1]

Fig. 1 Excerpt from Nature 424, 870–872 (21 August 2003)

Volume 111, Issue 7, p923–925, 27 December 2002

MINIREVIEW

Taking the Study of Cancer Cell Survival to a New Dimension

Tyler Jacks, Robert A. Weinberg

Fig. 2 Excerpt from Cell 111(7), 923–925 (27 December 2002)

This is made very poignant in a publication in *Cell* [43] in 2002 about the "new" dimension (third) in cell culture that had been around for almost the entire century of the history of tissue culture.

Meanwhile, Hans Clevers started growing his "organoids" on Matrigel and produced structures that he claimed resembled intestines and many other organs [36], originating from "stem cells" and stating that this is the only definition of "organoids." Using this strategy, Clevers has wrested leadership of the "3D culture" field from Mina, now in her middle 70s. Hans has been elected as a Foreign Associate of the U.S. National Academy of Sciences and is clearly aiming for a Nobel Prize for his "organoid" work.

Unknown to Mina, Hans, and probably Hynda, and unfortunately, to most workers in the field, is that Joseph Leighton invented 3D sponge-matrix histoculture almost 70 years ago, the theme of this volume. Sponge matrices are flexible and let cells and tissue take their natural shape, which Folkman understood 40 years ago is critical for proper gene expression and cell function [46]. Sponge matrices, such as Gelfoam®, which is gelatinized pig skin, are very rich in collagen and have many structures and interstices for cells and tissue to attach and invade, and yet are flexible allowing cells, organoids, and tissues to take their natural shape, which appears to be essential for in vivo-like gene expression [47]. Matrigel, in contrast to Gelfoam®, is a coating, does not have a structure, and is not flexible when coated on plastic. In this volume, Tome et al. [48] (Chapter 19) show that cancer cells grown on Gelfoam® can form complex structures, in contrast to the same cells grown on Matrigel, which only form aggregates (spheres). Therefore, the dominant form of cell culture today, Matrigel culture, is not really 3D, and we call it 2.5D [49].

This volume begins with the story of Joe Leighton, his invention of true 3D sponge-matrix histoculture, which enabled self-organization of organoids and tissue, cancer growth and invasion, and growth of normal and malignant tissue-like structure from adults and embryos in vitro. This volume describes many applications of sponge-matrix culture for the study of cancer, the treatment of cancer, the study of stem cells, growth of nerves and nerve repair, the study of lymphoid tissues that produce antibodies and are susceptible to HIV infection, and much more. Despite these very big advances, sponge-matrix culture is not the current paradigm.

Sponge-matrix histoculture has also been developed as a cancer-patient drug-response assay, called the histoculture drug-response assay (HDRA), for all tumor types, that improves the outcome of chemotherapy for patients (please see Chapters 5, 7, 8–11 in this volume). However, the *Journal of Clinical Oncology*, the main organ for medical oncologists worldwide, publishes paper after paper stating that such assays are "experimental" and not to be used in clinical practice [50, 51]. As mentioned above, science does not move in a straight line, and, as Kuhn stated, sometimes scientific paradigms can only be overturned when the "old guard" finally dies off [52].

It is the hope that this volume will contribute to a new dominant paradigm of cell and tissue culture for research on cancer and other diseases that is based on true 3D culture. Please enjoy the book, and I hope it will inspire you to join the field of true 3D culture.

San Diego, CA, USA   *Robert M. Hoffman*

References

1. Hoffman RM (2013) Tissue culture. In: Brenner's encyclopedia of genetics, 2nd edn, vol 7. Elsevier, p 73–76

2. Harrison RG (1970) Observations on the living developing nerve fiber. Proc Soc Exp Biol Med 4:140–143

3. Carrel A, Burrows MT (1910) Cultivation of adult tissues and organs outside of the body. J Am Med Assoc 55:1379–1381

4. Carrel A, Burrows MT (1911) Cultivation in vitro of malignant tumors. J Exp Med 13:571–575

5. Carrel A (1912) On the permanent life of tissues outside of the organism. J Exp Med 15:516–528

6. Rous P (1911) A sarcoma of the fowl transmissible by an agent separable from the tumor cells. J Exp Med 13:397–411

7. Carrel A (1923) A method for the physiological study of tissue in vitro. J Exp Med 38:407–418

8. Rous P, Jones FS (1916) A method for obtaining suspensions of living cells from fixed tissues and for the plating out of individual cells. J Exp Med 23:549–555

9. Earle WR, Schilling EL, Stark TH, Straus NP, Brown MF, Shelton E (1943) Production of malignancy in vitro. IV. The mouse fibroblast cultures and changes seen in living cells. J Natl Inst Cancer 4:165–212

10. Gey GO, Coffman WD, Kubicek MT (1954) Tissue culture studies on the proliferative capacity of cervical carcinoma and normal epithelium. Cancer Res 12:264–265

11. Ham RG (1965) Clonal growth of mammalian cells in a chemically defined synthetic medium. Proc Natl Acad Sci U S A 53:288–293

12. Eagle H (1955) Nutrition needs of mammalian cells in tissue culture. Science 122:501–504

13. Sanford KK, Earle WR, Likely GD (1948) The growth in vitro of single isolated tissue cells. J Natl Cancer Inst 9:229–246

14. Puck TT, Marcus PI (1955) A rapid method for viable cell titration and clone production with HeLa cells in tissue culture: the use of X-irradiated cells to supply conditioning factors. Proc Natl Acad Sci U S A 41:432–437

15. Puck TT, Marcus PI (1956) Action of X-rays on mammalian cells. J Exp Med 103:653–666

16. Harris H (1970) Cell fusion, the Dunham Lectures. Harvard University Press, Cambridge, MA

17. Ephrussi B (1972) Hybridization of somatic cells. Princeton University Press, Princeton, NJ

18. Enders JF, Robbins FC, Weller Th (1980) Classics in infectious diseases. The cultivation of the poliomyelitis viruses in tissue culture by John F. Enders, Frederick C. Robbins and Thomas H. Weller. Rev Infect Dis 2:493–504

19. Leighton J (1960) The propagation of aggregates of cancer cells: implications for therapy and a simple method of study. Cancer Chemother Rep 9:71–72

20. Leighton J, Kalla Rl, Turner JM Jr, Fennell RH Jr (1960) Pathogenesis of tumor invasion. II. Aggregate replication. Cancer Res 20:575–586

21. Leighton J (1959) Aggregate replication, a factor in the growth of cancer. Science 129(3347):466–467

22. Leighton J, Kalla Rl, Kline I, Belkin M (1959) Pathogenesis of tumor invasion. I. Interaction between normal tissues and transformed cells in tissue culture. Cancer Res 19:23–27

23. Dawe CJ, Potter M, Leighton J (1958) Progressions of a reticulum-cell sarcoma of the mouse in vivo and in vitro. J Natl Cancer Inst 21:753–781

24. Leighton J (1957) Contributions of tissue culture studies to an understanding of the biology of cancer: a review. Cancer Res 17:929–941

25. Kline I, Leighton J, Belkin M, Orr HC (1957) Some observations on the response of four established human cell strains to hydrocortisone in tissue culture. Cancer Res 17:780–784

26. Leighton J, Kline I, Belkin M, Legallais F, Orr HC (1957) The similarity in histologic appearance of some human cancer and normal cell strains in sponge-matrix tissue culture. Cancer Res 17:359–363

27. Leighton J, Kline I, Belkin M, Orr HC (1957) Effects of a podophyllotoxin derivative on tissue culture systems in which human cancer invades normal tissue. Cancer Res 17:336–344

28. Leighton J, Kline I, Belkin M, Tetenbaum Z (1956) Studies on human cancer using sponge-matrix tissue culture. III. The invasive properties of a carcinoma (strain HeLa) as influenced by temperature

variations, by conditioned media, and in contact with rapidly growing chick embryonic tissue. J Natl Cancer Inst 16:1353–1373

29. Leighton J, Kline I, Orr HC (1956) Transformation of normal human fibroblasts into histologically malignant tissue in vitro. Science 123:502

30. Leighton J (1954) The growth patterns of some transplantable animal tumors in sponge matrix tissue culture. J Natl Cancer Inst 15:275–293

31. Leighton J, Kline I (1954) Studies on human cancer using sponge matrix tissue culture. II. Invasion of connective tissue by carcinoma (strain HeLa). Tex Rep Biol Med 12:865–873

32. Leighton J (1954) Studies on human cancer using sponge matrix tissue culture. I. The growth patterns of a malignant melanoma, adenocarcinoma of the parotid gland, papillary adenocarcinoma of the thyroid gland, adenocarcinoma of the pancreas, and epidermoid carcinoma of the uterine cervix (Gey's HeLa strain). Tex Rep Biol Med 12:847–864

33. Leighton J (1951) A sponge matrix method for tissue culture; formation of organized aggregates of cells in vitro. J Natl Cancer Inst 12:545–561

34. Bissell MJ (2017) Goodbye flat biology—time for the 3rd and the 4th dimensions. J Cell Sci 130:3–5

35. Muthuswamy SK (2017) Bringing together the organoid field: from early beginnings to the road ahead. Development 144:963–967

36. Clevers H (2016) Modeling Development and Disease with Organoids. Cell 165(7):1586–1597

37. Thomson JA, Itskovitz-Eldor J, Shapiro SS, Waknitz MA, Swiergiel JJ, Marshall VS, Jones JM (1998) Embryonic stem cell lines derived from human blastocysts. Science 282:1145–1147

38. Yamanaka S (2013) The winding road to pluripotency (Nobel Lecture). Angew Chem Int Ed Engl 52:13900–13909

39. Kleinman HK, McGarvey ML, Hassell JR, Star VL, Cannon FB, Laurie GW, Martin GR (1986) Basement membrane complexes with biological activity. Biochemistry 25:312–318

40. Orkin RW, Gehron P, McGoodwin EB, Martin GR, Valentine T, Swarm R (1977) A murine tumor producing a matrix of basement membrane. J Exp Med 145:204–220

41. Weaver VM, Petersen OW, Wang F, Larabell CA, Briand P, Damsky C, Bissell MJ (1997) Reversion of the malignant phenotype of human breast cells in three-dimensional culture and in vivo by integrin blocking antibodies. J Cell Biol 137:231–245

42. Weaver VM, Lelièvre S, Lakins JN, Chrenek MA, Jones JC, Giancotti F, Werb Z, Bissell MJ (2002) beta4 integrin-dependent formation of polarized three-dimensional architecture confers resistance to apoptosis in normal and malignant mammary epithelium. Cancer Cell 2:205–216

43. Jacks T, Weinberg RA (2002) Taking the study of cancer cell survival to a new dimension. Cell 111:923–925

44. Abbott A (2003) Cell culture: Biology's new dimension. Nature 424:870–872

45. Simian M, Bissell MJ (2017) Organoids: a historical perspective of thinking in three dimensions. J Cell Biol 216:31–40

46. Folkman J, Moscona A (1978) Role of cell shape in growth control. Nature 273:345–349

47. Folkman J, Hochberg M (1973) Self-regulation of growth in three dimensions. J Exp Med 138:745–753

48. Tome Y, Uehara F, Mii S, Yano S, Zhang L, Sugimoto N, Maehara H, Bouvet M, Tsuchiya H, Kanaya F, Hoffman RM (2014) 3-dimensional tissue is formed from cancer cells in vitro on Gelfoam®, but not on Matrigel™. J Cell Biochem 115:1362–1367

49. Zhang L, Wu C, Bouvet M, Yano S, Hoffman RM (2015) Traditional Chinese medicine herbal mixture LQ arrests FUCCI-expressing HeLa cells in G0/G1 phase in 2D plastic, 2.5D Matrigel®, and 3D Gelfoam® culture visualized with FUCCI imaging. Oncotarget 6:5292–5298

50. Samson DJ, Seidenfeld J, Ziegler K, Aronson N (2004) Chemotherapy sensitivity and resistance assays: a systematic review. J Clin Oncol 22:3618–3630

51. Burstein HJ, Mangu PB, Somerfield MR, Schrag D, Samson D, Holt L, Zelman D, Ajani JA, American Society of Clinical Oncology (2011) American Society of Clinical Oncology clinical practice guideline update on the use of chemotherapy sensitivity and resistance assays. J Clin Oncol 29:3328–3330

52. Kuhn T (1962) The structure of scientific revolutions. University of Chicago Press, Chicago

Contents

Contributors

YASUYUKI AMOH • *Department of Dermatology, Kitasato University School of Medicine, Minami Ward, Sagihara, Japan*

WENLUO CAO • *AntiCancer Inc., San Diego, CA, USA; University of California, San Diego, CA, USA; Department of Anatomy, Second Military Medical University, Shanghai, China*

TAKASHI CHISHIMA • *Yokohama Rosai Hospital, Yokohama City, Kanagawa Prefecture, Japan*

JENNIFER DUONG • *AntiCancer Inc., San Diego, CA, USA*

WENDY FITZGERALD • *National Institute of Child Health and Human Development, National Institutes of Health, Bethesda, MD, USA*

AARON E. FREEMAN • *AntiCancer, Inc., San Diego, CA, USA*

FIORELLA GUADAGNI • *San Raffaele Roma Open University, Rome, Italy; IRCCS San Raffaele Pisana, Rome, Italy*

ROBERT M. HOFFMAN • *AntiCancer Inc., San Diego, CA, USA; Department of Surgery, UCSD, San Diego, CA, USA*

ANDREA INTROINI • *Karolinska Institute, Stockholm, Sweden*

PHILL-SEUNG JUNG • *Department of Obstetrics and Gynecology, University of Ulsan College of Medicine, Asan Medical Center, Seoul, South Korea*

FUMINORI KANAYA • *Department of Orthopedic Surgery, Graduate School of Medicine, University of the Ryukyus, Okinawa, Japan*

MOON-BO KIM • *MetaBio Institute, Seoul, South Korea*

LINGNA LI • *AntiCancer Inc., San Diego, CA, USA*

FANG LIU • *Department of Anatomy, Regenerative Medicine Center, Second Military Medical University, Shanghai, China*

LEONID MARGOLIS • *National Institute of Child Health and Human Development, National Institutes of Health, Bethesda, MD, USA*

SUMIYUKI MII • *AntiCancer Inc., San Diego, CA, USA; Department of Dermatology, Kitasato University School of Medicine, Minami Ward, Sagihara, Japan*

SHINJI MIWA • *Department of Orthopaedic Surgery, Kanazawa University Graduate School of Medicine, Kanazawa, Japan*

JOO-HYUN NAM • *Department of Obstetrics and Gynecology, University of Ulsan College of Medicine, Asan Medical Center, Seoul, South Korea*

HIROKAZU TANINO • *Department of Breast Surgery, Kobe University Hospital, Kobe, Japan*

YASUNORI TOME • *AntiCancer Inc., San Diego, CA, USA; Department of Surgery, University of California, San Diego, CA, USA; Department of Orthopedic Surgery, Graduate School of Medicine, University of the Ryukyus, Okinawa, Japan*

FUMINARI UEHARA • *AntiCancer Inc., San Diego, CA, USA; Department of Surgery, University of California, San Diego, CA, USA; Department of Orthopedic Surgery, Graduate School of Medicine, University of the Ryukyus, Okinawa, Japan*

CHRISTOPHE VANPOUILLE • *National Institute of Child Health and Human Development, National Institutes of Health, Bethesda, MD, USA*

ROBERT A. VESCIO • *Multiple Myeloma and Amyloidosis Program, Samuel Oschin Comprehensive Cancer Institute, Cedars-Sinai Medical Center, Los Angeles, CA, USA*

SHUYA YANO • *Department of Gastroenterological Surgery, Graduate School of Medicine, Dentistry and Pharmaceutical Sciences, Okayama University, Okayama, Japan*

Chapter 1

In Memoriam: Joseph Leighton, 1921–1999—Father of 3-Dimensional Tissue Culture

Robert M. Hoffman

Abstract

Joseph Leighton developed sponge-matrix histoculture in the early 1950s and is the father of true 3D cell and tissue culture. Dr. Leighton demonstrated that sponge-matrix histoculture could replicate in vivo-like tissue structure and function by allowing cells to take their natural shape.

Key words Sponge-matrix, 3-Dimensional, Histoculture, Tumors, Cell-self-organization, Organoid, Tissue

1 Joseph Leighton, Circa 1982

"The biology of animal cells in culture is often studied in individual cells or in sheets of cells. The relevance of such studies to the intact animal is unclear, since the spatial conditions encountered by cells in animals is one of dense three-dimensional masses of cells, with limits to migration, and with gradients of both diffusion of metabolites and of morphologic maturation"—Joseph Leighton, circa 1951 [1].

Robert M. Hoffman (ed.), *3D Sponge-Matrix Histoculture: Methods and Protocols*, Methods in Molecular Biology, vol. 1760, https://doi.org/10.1007/978-1-4939-7745-1_1, © Springer Science+Business Media, LLC, part of Springer Nature 2018

"Young investigators today often are not aware of prior work and may, at times, even see earlier work of their own either repeated or not even acknowledged"—Joseph Leighton, circa 1951 [1].

These long-ago quotes by Dr. Leighton are prophetic and highly relevant today in the very confused present-day field of "3D cell culture." Dr. Leighton invented a method in the early 1950s in which cells, cell aggregates, organoids, and tissues grew on and in a sponge gel and produced a basic tissue pattern similar to that observed in vivo. This method is called sponge-matrix culture [2].

In early studies, Leighton showed that cellular outgrowth of tumor explants in a sponge-matrix culture system had multiple areas of organized aggregates of cells that were very similar to the original tumor. Many glandular structures with distinct lumina were observed along with smaller acinous clusters of cells in the sponge matrix (Figs. 1, 2, and 3) [2].

Mammary adenocarcinoma, ovarian cancer, teratoma, malignant melanoma, reticulum cell sarcoma, lymphosarcoma and L sarcoma of the mouse, and ovarian papillary adenocarcinoma and hepatoma of the rat were grown in the sponge-matrix histoculture by Dr. Leighton. Highly organized outgrowths of the tumors were observed in the sponge matrix (Figs. 2, 3, and 4) [4].

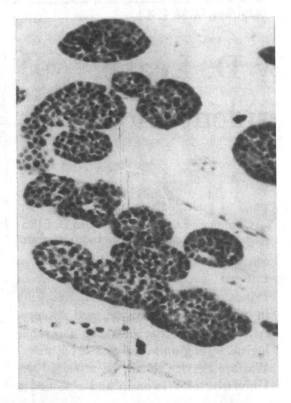

Fig. 1 Histologic section of sponge matrix culture of cancer cells. Aggregates gave rise to secondary aggregates [3]

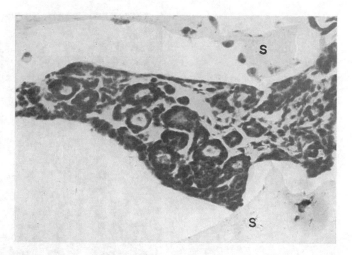

Fig. 2 Mouse mammary adenocarcinoma growing in a sponge matrix. The cellular aggregate, consisting largely of definite glandular structures, is covered by epithelium. Above and below the tissue, the sponge (marked "S") appears faintly gray with fine irregular stippling. Hematoxylin and eosin. ×400 [2]

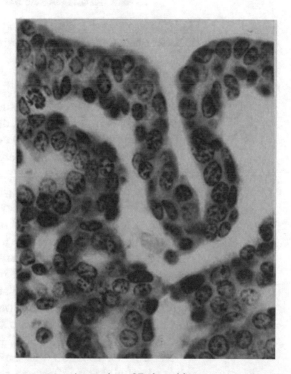

Fig. 3 Mouse mammary tumor in a 37-day-old sponge matrix culture in vitro. Detail of papillary glandular structure which shows a regular double-cell layer both on the outer surface and within the structure. Mitoses are numerous. Hematoxylin and eosin. ×S00 [4]

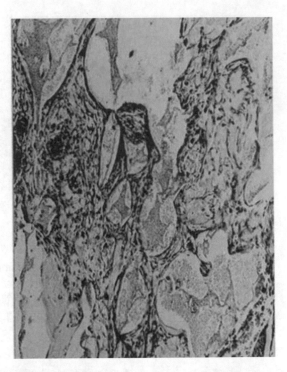

Fig. 4 Teratoma after 5 days in sponge-matrix culture. The tumor is growing within the sponge interstices. Hematoxylin and eosin. ×100 [4]

Leighton observed that HeLa human carcinoma cells invaded normal human connective tissue derived from human foreskin fibroblasts in sponge-matrix culture (Fig. 5) [6]. The invading HeLa cells replaced the normal cells in the connective tissue [5].

Chick embryonic tissues were cultured by Dr. Leighton along with HeLa cells on sponge-matrices. The mesenchyme-like embryonic connective tissues from the heart and bone tissue were infiltrated by HeLa cells in the Gelfoam® sponge matrix. Embryonic intestine and lung invaded the HeLa cells in the Gelfoam® sponge-matrix (Figs. 6, 7, 8, and 9) [5].

Leighton grew various normal human tissues including connective tissue in sponge-matrix tissue culture along with different types of cancer cells. Most cellular infiltration was found where the alignment of connective tissue was at right angles to the margin of the growing cancer cells where connective tissue and cancer cells infiltrated each other (Fig. 5) [7].

An infant's foreskin was explanted into sponge-matrix tissue culture by Leighton [7] where the outgrowing cells formed a lattice of spindle-shaped cells and collagenous fibers, forming connective tissue throughout the sponge interstices. The foreskin fibroblasts also produce a dense layer of spindle-shaped cells on the outer surfaces of the sponge [7]. A 15th subculture had large unusual cells extending from the edges of the sponge onto the

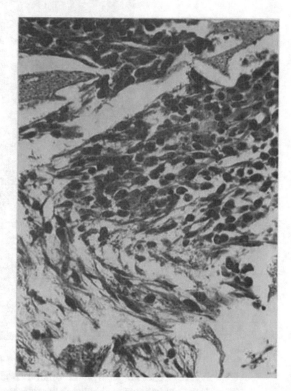

Fig. 5 HeLa cells and connective tissue cells in sponge-matrix culture are intermingled in the invading margin of HeLa cells. Hematoxylin and eosin. ×250 [5]

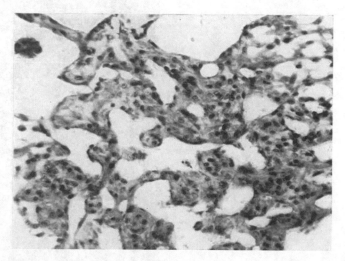

Fig. 6 Mixed chick embryo tissues in sponge matrix culture showing endothelial-lined channels derived from chick-embryo heart. Hematoxylin and eosin. ×400 [2]

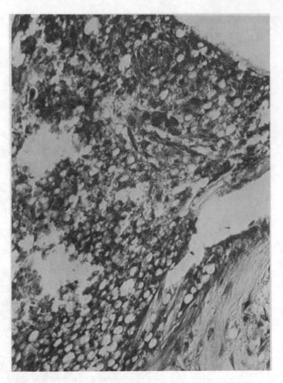

Fig. 7 Sponge-matrix culture of tissue of the chick embryonic liver invaded by HeLa cells. Large, hyperchromatic cancer cells are seen among many smaller, vacuolated liver epithelial cells. Hematoxylin and eosin. ×250 [5]

Fig. 8 Sponge matrix culture of tissue of the chick embryonic brain along with HeLa cells. HeLa cells are in the upper half of the field among the embryonic brain cells. Hematoxylin and eosin. ×250 [5]

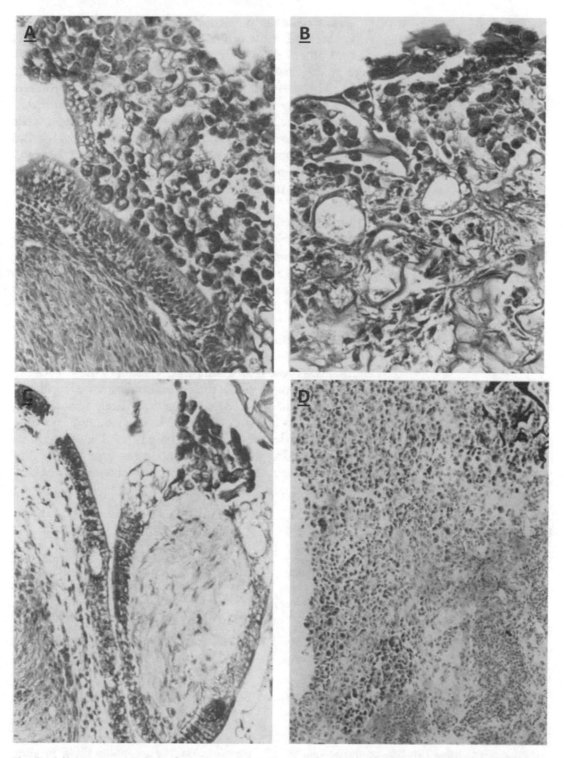

Fig. 9 (**a**) HeLa cells in Gelfoam® sponge matrix culture along with embryonic chick intestinal epithelium. (**b**) An area of the Gelfoam® explant of A, in which a cyst lined by embryonic intestinal epithelium of the chick is seen deep in the HeLa-Gelfoam® inoculum. (**c**) An area in which HeLa cells are in contact with degenerating, sparse cellular connective tissue of intestine from the chick. (**d**) HeLa and embryonic lung of the chick in contact [5]

adjacent glass wall of the tube containing dense cellular tissue and large hyperchromatic cells with large nuclei containing enormous nucleoli, and other cells with four to six nuclei. Many mitotic figures were multipolar resembling a highly malignant anaplastic tumor [8]. These results suggest that the foreskin fibroblasts underwent a malignant transformation in sponge-matrix culture.

The sponge-matrix histoculture systems with cancer cells invading normal tissue were used by Leighton to determine the efficacy of experimental chemotherapeutic agents, such as acetyl podophyllotoxin-to-pyridinium chloride (NCI-3022). A dose of 1 μg/mL inhibited the cancer cells much more than the normal tissue [9].

Leighton noted that human cancer cells in sponge-matrix tissue culture produced cell aggregates. Large confluent masses of cancer cells were observed through contiguous interstices in the sponge. Aggregate multiplication, in vitro similar to what occurs in vivo, was observed. Small aggregates separated from a parent aggregate (Fig. 1) [10].

Leighton noted that hyperplastic liver nodules from monolayer culture had a much more organoid arrangement with duct-like structures in sponge-matrix histoculture [11].

Leighton was more than a half century ahead of the field of 3D cell and tissue culture, in which the current mainstream is still well behind Dr. Leighton [12, 13] and has completely ignored him (please see the Foreword and Afterword in this volume).

Thiss volume describes studies based on Dr. Leighton's momentous pioneering work demonstrating the great potential of sponge-matrix histoculture to represent in vivo conditions of cancer and normal tissue in vitro.

References

1. Schaeffer WI (1999) In memorium—a tribute to Dr. Joseph Leighton. Methods Cell Sci 21:v–vi

2. Leighton J (1951) A sponge matrix method for tissue culture; formation of organized aggregates of cells in vitro. J Natl Cancer Inst 12:545–561

3. Leighton J (1959) Aggregate replication, a factor in the growth of cancer. Science 129:466–467

4. Leighton J (1954) The growth patterns of some transplantable animal tumors in sponge matrix tissue culture. J Natl Cancer Inst 15:275–293

5. Leighton J, Kline I, Belkin M, Tetenbaum Z (1956) Studies on human cancer using sponge-matrix tissue culture. III. The invasive properties of a carcinoma (strain HeLa) as influenced by temperature variations, by conditioned media, and in contact with rapidly growing chick embryonic tissue. J Natl Cancer Inst 16:1353–1373

6. Leighton J, Kline I (1954) Studies on human cancer using sponge matrix tissue culture. II. Invasion of connective tissue by carcinoma (strain HeLa). Tex Rep Biol Med 12:865–873

7. Leighton J, Kalla R, Kline I, Belkin M (1959) Pathogenesis of tumor invasion. I. Interaction between normal tissues and transformed cells in tissue culture. Cancer Res 19:23–27

8. Leighton J, Kline I, Hc O (1956) Transformation of normal human fibroblasts into histologically malignant tissue in vitro. Science 123:502–503

9. Leighton J, Kline I, Belkin M, Hc O (1957) Effects of a podophyllotoxin derivative on tissue

culture systems in which human cancer invades normal tissue. Cancer Res 17:336–344

10. Leighton J (1957) Contributions of tissue culture studies to an understanding of the biology of cancer: a review. Cancer Res 17:929–941

11. Slifkin M, Merkow LP, Pardo M, Epstein SM, Leighton J, Farber E (1970) Growth in vitro of cells from hyperplastic nodules of liver induced by 2-fluorenylacetamide or aflatoxin B1. Science 167:285–287

12. Leighton J (1969) Propagation of cancer: targets for future chemotherapy. Cancer Res 29:2457–2465

13. Leighton J, Mark R, Justh G (1968) Patterns of three-dimensional growth in vitro in collagen-coated cellulose sponge: carcinomas and embryonic tissues. Cancer Res 28:286–296

Chapter 2

3D Sponge-Matrix Histoculture: An Overview

Robert M. Hoffman

Abstract

Three-dimensional cell culture and tissue culture (histoculture) is much more in vivo-like than 2D culture on plastic. Three-dimensional culture allows investigation of crucial events in tumor biology such as drug response, proliferation and cell cycle progression, cancer cell migration, invasion, metastasis, immune response, and antigen expression that mimic in vivo conditions. Three-dimensional sponge-matrix histoculture maintains the in vivo phenotype, including the formation of differentiated structures of normal and malignant tissues, perhaps due to cells maintaining their natural shape in a sponge-gel matrix such as Gelfoam®. Sponge-matrix histoculture can also support normal tissues and their function including antibody-producing lymphoid tissue that allows efficient HIV infection, hair-growing skin, excised hair follicles that grow hair, pluripotent stem cells that form nerves, and much more.

Key words Gelfoam®, 3D, Histoculture, Tumors, Skin, Hair growth, Tonsils, Antibodies, Infection, HIV, Stem cells

1 Introduction

1.1 Three-Dimensional Culture Is In Vivo-Like in Contrast to 2D Culture

Folkman and Moscona [1] observed that cell shape greatly influenced gene expression. This may be a key point why 3D culture is much more in vivo-like compared to monolayer culture where cells are forced to take a very unnatural shape. The most striking difference between 2D and 3D culture is drug sensitivity, where 3D cultures may be 3–4 orders of magnitude more resistant to the same drug than 2D culture of the same cells [2].

For example, Heppner and colleagues [3, 4] compared cancer-cell drug response in 2D and 3D culture. They found that drug resistance in the 3D cultures could be up to 1000-fold greater than in monolayer cultures. In 3D bolus cultures in collagen, the cells grew even in the presence of drug concentrations that reduced survival in monolayers to less than 0.1% of controls. When the cells from the collagen gels were replated as monolayers, they became sensitive again. Three-dimensional structure itself rather than just simple inaccessibility to nutrients allowed for the increased drug resistance. For example, cells in monolayer cultures were initially exposed to melphalan and

Robert M. Hoffman (ed.), *3D Sponge-Matrix Histoculture: Methods and Protocols*, Methods in Molecular Biology, vol. 1760, https://doi.org/10.1007/978-1-4939-7745-1_2, © Springer Science+Business Media, LLC, part of Springer Nature 2018

5-fluorouracil (5-FU) by Heppner and colleagues. The cultures were transferred to collagen gels and the cells became highly resistant to these drugs. Even though the cells were exposed to the drugs as monolayers, where the drugs can access the cells readily and diffusion is not limiting, when the cells were transferred after drug exposure to a 3D structure, they were resistant to the drugs. It is possible that in 3D culture the cancer cells entered G_1/G_0 and thereby became resistant to the drugs [5].

Another example of the importance of 3D culture was demonstrated when the EMT-6 tumor in mice was treated with chemotherapy in order to select highly resistant variant in vivo [6]. When cells from these tumors were grown as monolayer cultures, they became as drug sensitive as the parental cells. It was mistakenly concluded that certain types of drug sensitivity may only be expressed in vivo. However, when the drug-resistant cancer cell lines were grown as spheroids in 3D cultures, resistance was observed up to almost 5000 times that of the parent with certain drugs. This resistance was not observed in monolayer culture, even when the cancer cells were cultured on laminin or Matrigel™ [7] demonstrating that Matrigel is not true 3D culture (please *see* Foreword, Chapter 19, and Afterword in the present volume). Cancer cells resistant to supra-pharmacological doses of drugs were able to form compact spheroids [7]. It is possible that the compact, drug-resistant spheroids have a large fraction of cells in the G_1 phase of the cell cycle due to the limited availability of nutrients and oxygen. G_1 cell cycle arrest would confer drug resistance [8–14].

The migration of aggregated cells in histoculture suggested to Leighton [15–29] (please *see* Chapter 1 in the present volume) that aggregates were the seeds of metastasis rather than single cells. It was later observed that the EMT-6 cells selected to be highly resistant in vivo, thereby forming compact cell aggregates, are also highly metastatic [30]. Drug resistance may therefore promote metastasis because it promotes aggregation. Thus, the formation of highly aggregated emboli due to acquisition of drug resistance may promote metastasis in the patient [17].

When cells were transferred from monolayer to 3D culture, there was a strong upregulation (up to 15-fold) of p27Kip1 protein in spheroids of human breast, ovarian, and colorectal carcinoma. Antisense-mediated downregulation of p27Kip1 in EMT-6 mammary tumor cell spheroids reduced intercellular adhesion, increased cell proliferation, and sensitized cancer cells to an activated form of cyclophosphamide. These results implicate $p27^{Kip1}$, possibly due to cell cycle arrest in G_1 [8–14], as a major regulator of the drug resistance of solid tumors [30].

1.2 Use of Gelfoam® Histoculture for Individualized Drug Response Assays

A major clinical problem is that cancers that are classified as identical according to their histopathological characteristics are nonetheless highly individual in their drug sensitivities and there is currently no way to *a priori* predict the clinical outcome of chemotherapy for individual patients [31]. Sponge-matrix histoculture was originally developed for growth of human tumors [15–29] (please *see* Chapter 1 of the present volume) and is a general system that enables growth of tumors at high frequency directly from surgery or biopsy that maintains important in vivo properties in vitro, including drug-response patterns [32].

An important example of the power of Gelfoam® histoculture of tumors is their use in individual patient drug-response testing. This was shown in a series of histocultured tumors of the same histopathological type from different individuals which were shown to be very different in their sensitivities to a single drug. The range of sensitivity to of histocultured breast cancers to doxorubicin (DOX), for example, spanned orders of magnitude, which reflects the clinical situation for breast cancer patients in general (please *see* Chapters 5, 7, 8 and 9 of the present volume describing clinical correlation of the Gelfoam® histoculture drug-response assay [HDRA]) [32].

2 Other Applications of Gelfoam® Histoculture

2.1 Gelfoam® Histoculture Androgen Sensitivity Assay

Sponge-matrix Gelfoam® histoculture was used to assay androgen sensitivity in human benign and malignant prostatic tissue. Dihydrotestosterone (DHT)-treated cultures increased their incorporation of [³H]thymidine. The degree of determined androgen sensitivity of individual patients with prostate cancer or benign prostate hypertrophy (BPH) [33] was in Gelfoam® histoculture (please *see* Chapter 11 of the present volume).

2.2 Gelfoam® Histoculture Skin Toxicity Assay

Sponge-matrix histoculture allows for the testing of toxicity of intact skin that contains hair follicles, which probably have the most chemo-sensitive cells in the body [34, 35]. In sponge-matrix Gelfoam® histoculture, all the cell types of skin remain viable and maintain their native architecture for at least 10 days. The culture system can be used for toxicity measurements by ascertaining cell viability using fluorescent dyes specific for living cells, and propidium iodide, specific for dead cells. The hair follicle cells were the most sensitive cells to DOX [34] (please *see* Chapter 14 of the present volume).

2.3 In Vitro Hair Growth in Gelfoam® Histoculture

Human scalp skin, with abundant hair follicles in various stages of the hair growth cycle, was grown for up to 40 days in sponge-matrix Gelfoam® histoculture at the air/liquid interface. Hair shaft elongation occurred mainly during the first 10 days of histoculture of both intact skin and isolated hair follicles. However, hair follicles were viable and follicle keratinocytes continued to incorporate [³H] thymidine for up to several weeks after hair shaft elongation ceased.

Spontaneous catagen was observed both histologically and by the actual regression of the hair follicle in Gelfoam® histoculture. In addition, vellus follicles were shown to be viable at day 40 after initiation of culture [36] (please *see* Chapter 14 of the present volume).

In subsequent studies, transgenic mice in which the nestin promoter drives green fluorescent protein (ND-GFP) were used as the source of the whiskers, allowing imaging of the nestin-expressing hair follicle-associated pluripotent (HAP) stem cells during hair growth. We showed that Gelfoam®-histocultured whisker follicles produced growing pigmented and unpigmented hair shafts. Hair-shaft length increased rapidly by day 4 and continued growing until at least day 12 after which the hair-shaft length was constant. By day 63 in histoculture, the number of ND-GFP-expressing HAP stem cells increased significantly in the follicles [37] (please *see* Chapter 14 of the present volume).

2.4 Hair Follicle-Associated Pluripotent (HAP) Stem Cells Form Nerves in Gelfoam® Histoculture

Vibrissa hair follicles, including their sensory nerve stump, were excised from ND-GFP mice and were placed in Gelfoam® histoculture. β-III tubulin-positive fibers, consisting of ND-GFP-expressing cells, extended up to 500 μm from the whisker nerve stump in Gelfoam® histoculture over time. The growing fibers had growth cones on their tips expressing F-actin. These findings indicate that β-III tubulin-positive fibers elongating from the whisker follicle sensory nerve stump were growing axons. The growing whisker sensory nerve was highly enriched in ND-GFP cells which appeared to play a major role in its elongation and interaction with other nerves in Gelfoam® histoculture, including the sciatic nerve, trigeminal nerve, and trigeminal nerve ganglion [38] (please *see* Chapter 16 of the present volume).

2.5 Gelfoam® Histoculture of Lymphoid Tissue

Human tonsils were grown in Gelfoam® histoculture and were viable for at least 4 weeks. Well-defined germinal centers of B cells, T cells, and macrophages were formed. Extended networks of follicular dendritic cells were found inside germinal centers [39]. Human immunodeficiency virus (HIV) was injected in histocultured tonsils. Efficient production of HIV in histocultures of human tonsil was observed. HIV production did not require stimulation by phytohemagglutinin (PHA) nor activation with interleukin-2 (IL-2) as did HIV infection of peripheral blood mononuclear cells (PRMCs) [39]. Histocultured lymphoid tissue produced specific antibodies upon challenge with (recall) antigens, such as diphtheria toxoid or tetanus toxoid [39] (please *see* Chapter 17 of the present volume).

2.6 Use of Green Fluorescent Protein (GFP) for Imaging of Tumor-Host Interaction in Gelfoam® Histoculture

Another advantage of histoculture is the ability to visualize in real time the development of metastasis using green fluorescent protein (GFP) imaging [40]. GFP-expressing human lung cancer cells and mouse lung were placed in Gelfoam® histoculture. Tumor progression was continuously visualized by GFP fluorescence in the same individual cultures over a 52-day period. The lung cancer cells spread throughout the mouse lung. Histocultured lung colonization was selective for lung cancer cells, because no lung cancer cell growth occurred on histocultured mouse liver tissue. The ability to support selective organ colonization in Gelfoam® histoculture and visualize tumor progression by GFP fluorescence allows the in vitro study of the governing processes of metastasis [41] (please *see* Chapter 20 of the present volume).

2.7 Cell Cycle Imaging in Gelfoam® Histoculture

A fluorescence ubiquitination-based cell cycle indicator (FUCCI) was used to image the phases of the cell cycle of cancer cells growing in Gelfoam® histoculture. In the FUCCI system, G_0/G_1 cells express a red fluorescent protein and $S/G_2/M$ cells express a green fluorescent protein. Cell cycle-phase distribution of cancer cells in Gelfoam® and in vivo tumors was similar. In both Gelfoam® histoculture and in vivo tumors, only the surface cells proliferated and interior cells were quiescent in G_0/G_1. This is in contrast to 2D culture where most cancer cells were always in cycle. The cancer cells responded similarly to toxic chemotherapy in Gelfoam® culture as in vivo, and very differently than cancer cells in 2D culture which were much more chemosensitive [42].

Gelfoam® histoculture and FUCCI imaging were used to determine the cell cycle position of invading and non-invading cells. With FUCCI 3D confocal imaging, we observed that cancer cells in G_0/G_1 phase in Gelfoam® histoculture migrated more rapidly and further than cancer cells in $S/G_2/M$ phases. Cancer cells ceased migrating when they entered $S/G_2/M$ phases and restarted migrating after cell division when the cells reentered G_0/G_1. Migrating cancer cells also were resistant to cytotoxic chemotherapy, since they were preponderantly in G_0/G_1, where cytotoxic chemotherapy is not effective [43] (please *see* Chapter 12 of the present volume).

2.8 Comparison of Gelfoam® Histoculture to Matrigel™ Culture to 2D Culture

Human osteosarcoma cells, stably expressing a fusion protein of α_v integrin and GFP, grew as a simple monolayer without any structure formation on the surface of a plastic dish. When the osteosarcoma cells were cultured within Matrigel™, the cancer cells formed colonies but no other structures. When the cancer cells were seeded on Gelfoam®, the cells formed 3D tissue-like structures. The behavior of osteosarcoma cells on Gelfoam® in culture was remarkably different from those cells in monolayer culture or in Matrigel™. Tissue-like structures were observed only in Gelfoam® culture confirming, as noted above, that Matrigel culture is not true 3D [44] (please *see* Chapter 19 of this volume).

References

1. Folkman J, Moscona A (1978) Role of cell shape in growth control. Nature 273:345–349

2. Hoffman RM (1993) To do tissue culture in two or three dimensions? That is the question. Stem Cells 11:105–111

3. Miller BE, Miller FR, Heppner GH (1984) Assessing tumour drug sensitivity by a new in vitro assay which preserves tumour heterogeneity and subpopulation interactions. J Cell Physiol 3(Suppl):105–116

4. Miller BE, Miller FR, Heppner GH (1985) Factors affecting growth and drug sensitivity of mouse mammary tumor lines in collagen gel cultures. Cancer Res 45:4200–4205

5. Yano S, Zhang Y, Miwa S, Tome Y, Hiroshima Y, Uehara F, Yamamoto M, Suetsugu A, Kishimoto H, Tazawa H, Zhao M, Bouvet M, Fujiwara T, Hoffman RM (2014) Spatial-temporal FUCCI imaging of each cell in a tumor demonstrates locational dependence of cell cycle dynamics and chemoresponsiveness. Cell Cycle 13:2110–2119

6. Teicher BA, Herman TS, Holden S, Wang YY, Pfeffer MR, Crawford JW, Frei E III (1990) Tumor resistance to alkylating agents conferred by mechanisms operative only in vivo. Science 247:1457–1461

7. Kobayashi HI, Man S, Graham C, Kapitain SJ, Teicher BA, Kerbel RS (1993) Acquired multicellular mediated resistance to alkylating agents in cancer. Proc Natl Acad Sci U S A 90:3294–3298

8. Yano S, Miwa S, Kishimoto H, Uehara F, Tazawa H, Toneri M, Hiroshima Y, Yamamoto M, Urata Y, Kagawa S, Bouvet M, Funiwara T, Hoffman RM (2015) Targeting tumors with a killer-reporter adenovirus for curative fluorescence-guided surgery of soft-tissue sarcoma. Oncotarget 6:13133–13148

9. Yano S, Miwa S, Kishimoto H, Toneri M, Hiroshima Y, Yamamoto M, Bouvet M, Urata Y, Tazawa H, Kagawa S, Funiwara T, Hoffman RM (2015) Experimental curative fluorescence-guided surgery of highly invasive glioblastoma multiforme selectively labeled with a killer-reporter adenovirus. Mol Ther 23:1182–1188

10. Yano S, Hiroshima Y, Maawy A, Kishimoto H, Suetsugu A, Miwa S, Toneri M, Yamamoto M, Katz MHG, Fleming JB, Urata Y, Tazawa H, Kagawa S, Bouvet M, Fujiwara T, Hoffman RM (2015) Color-coding cancer and stromal cells with genetic reporters in a patient-derived orthotopic xenograft (PDOX) model of pancreatic cancer enhances fluorescence-guided surgery. Cancer Gene Ther 22:344–350

11. Yano S, Zhang Y, Miwa S, Kishimoto H, Urata Y, Bouvet M, Kagawa S, Fujiwara T, Hoffman RM (2015) Precise navigation surgery of tumours in the lung in mouse models enabled by in situ fluorescence labelling with a killer-reporter adenovirus. BMJ Open Respir Res 2:e000096

12. Yano S, Takehara K, Miwa S, Kishimoto H, Hiroshima Y, Murakami T, Urata Y, Kagawa S, Bouvet M, Fujiwara T, Hoffman RM (2016) Improved resection and outcome of colon-cancer liver metastasis with fluorescence-guided surgery using in situ GFP labeling with a telomerase-dependent adenovirus in an orthotopic mouse model. PLoS One 11:e0148760

13. Yano S, Miwa S, Kishimoto H, Urata Y, Tazawa H, Kagawa S, Bouvet M, Fujiwara T, Hoffman RM (2016) Eradication of osteosarcoma by fluorescence-guided surgery with tumor labeling by a killer-reporter adenovirus. J Orthopaedic Res 34:836–844

14. Yano S, Takehara K, Miwa S, Kishimoto H, Tazawa H, Urata Y, Kagawa S, Bouvet M, Fujiwara T, Hoffman RM (2016) Fluorescence-guided surgery of a highly-metastatic variant of human triple-negative breast cancer targeted with a cancer-specific GFP adenovirus prevents recurrence. Oncotarget 7:75635–75647

15. Leighton J (1960) The propagation of aggregates of cancer cells: implications for therapy and a simple method of study. Cancer Chemother Rep 9:71–72

16. Leighton J, Kalla R, Turner JM Jr, Fennell RH Jr (1960) Pathogenesis of tumor invasion. II. Aggregate replication. Cancer Res 20:575–586

17. Leighton J (1959) Aggregate replication, a factor in the growth of cancer. Science 129(3347):466–467

18. Leighton J, Kalla R, Kline I, Belkin M (1959) Pathogenesis of tumor invasion. I. Interaction between normal tissues and transformed cells in tissue culture. Cancer Res 19:23–27

19. Dawe CJ, Potter M, Leighton J (1958) Progressions of a reticulum-cell sarcoma of the mouse in vivo and in vitro. J Natl Cancer Inst 21:753–781

20. Leighton J (1957) Contributions of tissue culture studies to an understanding of the biology of cancer: a review. Cancer Res 17:929–941

21. Kline I, Leighton J, Belkin M, Orr HC (1957) Some observations on the response of four established human cell strains to hydrocortisone in tissue culture. Cancer Res 17:780–784

22. Leighton J, Kline I, Belkin M, Legallais F, Orr HC (1957) The similarity in histologic appearance of some human cancer and normal cell

strains in sponge-matrix tissue culture. Cancer Res 17:359–363

23. Leighton J, Kline I, Belkin M, Orr HC (1957) Effects of a podophyllotoxin derivative on tissue culture systems in which human cancer invades normal tissue. Cancer Res 17:336–344

24. Leighton J, Kline I, Belkin M, Tetenbaum Z (1956) Studies on human cancer using sponge-matrix tissue culture. III. The invasive properties of a carcinoma (strain HeLa) as influenced by temperature variations, by conditioned media, and in contact with rapidly growing chick embryonic tissue. J Natl Cancer Inst 16:1353–1373

25. Leighton J, Kline I, Orr HC (1956) Transformation of normal human fibroblasts into histologically malignant tissue in vitro. Science 123:502

26. Leighton J (1954) The growth patterns of some transplantable animal tumors in sponge matrix tissue culture. J Natl Cancer Inst 15:275–293

27. Leighton J, Kline I (1954) Studies on human cancer using sponge matrix tissue culture. II. Invasion of connective tissue by carcinoma (strain HeLa). Tex Rep Biol Med 12:865–873

28. Leighton J (1954) Studies on human cancer using sponge matrix tissue culture. I. The growth patterns of a malignant melanoma, adenocarcinoma of the parotid gland, papillary adenocarcinoma of the thyroid gland, adenocarcinoma of the pancreas, and epidermoid carcinoma of the uterine cervix (Gey's HeLa strain). Tex Rep Biol Med 12:847–864

29. Leighton J (1951) A sponge matrix method for tissue culture; formation of organized aggregates of cells in vitro. J Natl Cancer Inst 12:545–561

30. St Croix B, Florenes VA, Rak JW, Flanagan M, Bhattacharya N, Slingerland JM, Kerbel RS (1996) Impact of the cyclin-dependent kinase inhibitor p27Kip1 on resistance of tumor cells to anticancer agents. Nat Med 2:1204–1210

31. Chabner BA (ed) (1983) Rational basis for chemotherapy. Liss, New York

32. Vescio RA, Redfern CH, Nelson TJ, Ugoretz S, Stern PH, Hoffman RM (1987) In vivo-like drug responses of human tumors growing in three-dimensional gel-supported, primary culture. Proc Natl Acad Sci U S A 84:5029–5033

33. Geller J, Sionit LR, Connors KM, Hoffman RM (1992) Measurement of androgen sensitivity in the human prostate in in vitro three-dimensional histoculture. Prostate 21:269–278

34. Li L, Margolis LB, Hoffman RM (1991) Skin toxicity determined in vitro by three-dimensional, native-state histoculture. Proc Natl Acad Sci U S A 88:1908–1912

35. Hoffman RM (2010) Histocultures and their use. In: Encyclopedia of life sciences. Wiley, Hoboken. (Published Online)

36. Li L, Margolis LB, Paus R, Hoffman RM (1992) Hair shaft elongation, follicle growth, and spontaneous regression in long-term, gelatin sponge-supported histoculture of human scalp skin. Proc Natl Acad Sci U S A 89:8764–8768

37. Cao W, Li L, Mii S, Amoh Y, Liu F, Hoffman RM (2015) Extensive hair shaft elongation by isolated mouse whisker follicles in very long-term Gelfoam® histoculture. PLoS One 10:e0138005

38. Mii S, Duong J, Tome Y, Uchugonova A, Liu F, Amoh Y, Saito N, Katsuoka K, Hoffman RM (2013) The role of hair follicle nestin-expressing stem cells during whisker sensory-nerve growth in long-term 3D culture. J Cell Biochem 114:1674–1684

39. Glushakova S, Baibakov B, Margolis LB, Zimmerberg J (1995) Infection of human tonsil histocultures: a model for HIV pathogenesis. Nature Med 1:1320–1322

40. Chishima T, Yang M, Miyagi Y, Li L, Tan Y, Baranov E, Shimada H, Moossa AR, Penman S, Hoffman RM (1997) Governing step of metastasis visualized in vitro. Proc Natl Acad Sci U S A 94:11573–11576

41. Chishima T, Miyagi Y, Li L, Tan Y, Baranov E, Yang M, Shimada H, Moossa AR, Hoffman RM (1997) Use of histoculture and green fluorescent protein to visualize tumor cell host interaction. In Vitro Cell Dev Biol 33:745–747

42. Yano S, Miwa S, Mii S, Hiroshima Y, Uehara F, Kishimoto H, Tazawa H, Zhao M, Bouvet M, Fujiwara T, Hoffman RM (2015) Cancer cells mimic in vivo spatial-temporal cell-cycle phase distribution and chemosensitivity in 3-dimensional Gelfoam® histoculture but not 2-dimensional culture as visualized with real-time FUCCI imaging. Cell Cycle 14:808–819

43. Yano S, Miwa S, Mii S, Hiroshima Y, Uehara F, Yamamoto M, Kishimoto H, Tazawa H, Bouvet M, Fujiwara T, Hoffman RM (2014) Invading cancer cells are predominantly in G_0/G_1 resulting in chemoresistance demonstrated by real-time FUCCI imaging. Cell Cycle 13:953–960

44. Tome Y, Uehara F, Mii S, Yano S, Zhang L, Sugimoto N, Maehara H, Bouvet M, Tsuchiya H, Kanaya F, Hoffman RM (2014) 3-dimensional tissue is formed from cancer cells in vitro on Gelfoam®, but not on Matrigel™. J Cell Biochem 115:1362–1367

In Vivo-Like Growth Patterns of Multiple Types of Tumors in Gelfoam® Histoculture

Robert M. Hoffman and Aaron E. Freeman

Abstract

Diverse human tumors obtained directly from surgery or biopsy can grow at high frequency in 3-dimensional Gelfoam® histoculture for long periods of time and still maintain many of their in vivo properties. The in vivo properties maintained in vitro include 3-dimensional growth; maintenance of tissue organization and structure, such as changes associated with oncogenic transformation; retention of differentiated function; tumorigenicity; and growth of multiple types of cells from a single tumor.

Key words Gelfoam® histoculture, Tumors, 3-Dimensional, Tissue architecture function

1 Introduction

This chapter reviews that human tumors can grow at high frequency for long periods of time in vitro and still maintain many of their critical in vivo properties. Cancer cells growing in monolayer culture or as clones in semisolid medium are, at best, only partly representative of the original tumor. Cells in monolayer culture have no chance to grow in three dimensions and to form organized tissues as would be the case in vivo. Clones of cancer cells in semi-solid medium may represent only a minority of the cells of the tumor, with the rest incapable of growing under the particular in vitro conditions provided. Cloning eliminates cell-cell interactions among heterogeneous cell types in tumors, which may be critical for the expression of many of their properties [1, 2]. Other workers (for example, refs. [3–6]) have stressed the importance of 3-dimensional in vitro growth of cancer cells in representing the in vivo situation. For a cultured tumor to be representative of actual cancer, it is essential that the tumor, as it proliferates in vitro, maintain its tissue organization and structure, its oncogenic properties, its differentiated functions, and any cellular heterogeneity that may have been present in vivo [7]. If human tumors growing in vitro can satisfy the above criteria and, in addition, can be grown

Robert M. Hoffman (ed.), *3D Sponge-Matrix Histoculture: Methods and Protocols*, Methods in Molecular Biology, vol. 1760,
https://doi.org/10.1007/978-1-4939-7745-1_3, © Springer Science+Business Media, LLC, part of Springer Nature 2018

at high frequency for long periods of time in culture, they should prove valuable for basic studies in cancer biology as well as for clinically-relevant testing. This chapter addresses the important fact that human tumors can indeed satisfy the above criteria in Gelfoam® histoculture [8].

2 Materials

2.1 Culture Medium [9]

1. Culture medium (CM): RPMI 1640 with L-glutamine and phenol red, 100 μM modified Eagle's medium (MEM)-nonessential amino acids (Thermo Fisher Scientific), 1 mM sodium pyruvate, 50 μg/mL gentamycin sulfate, 2.5 μg/mL amphotericin B (Thermo Fisher Scientific).

2. 10% Fetal bovine serum (FBS) (Gibco): We recommend testing different batches of FBS before purchasing a large stock.

3. 200 mM L-Glutamine (Gibco).

2.2 Gelfoam® Sponge [9]

1. Gelfoam® Absorbable Collagen Sponge USP, 12–7 mm format (Pfizer, NDC: 0009-031508).

2.3 Materials for Preparation of Collagen Sponges [9]

1. Gelfoam® trimming aid: Double-edged blade.

2. Sterile Petri dish, 100 mm × 20 mm or wider.

3. Tissue transportation medium: Sterile phosphate buffer saline (PBS) or other solution with physiologic pH.

4. Sterile transportation container.

5. Sterile metal forceps or tweezers.

6. Sterile metal scissors.

7. Sterile flat weighing metal spatula.

8. 6-Well culture plates.

9. Water-jacketed CO_2 incubator, set at 37 °C, 5% CO_2, ≥90% humidity.

10. Water bath.

2.4 Processing of Human Tumor Tissue [9]

1. Specimen of human tumors obtained from surgery.

2. Sterile Petri dish, 100 mm × 20 mm.

3. 70% ethanol solution.

4. Disinfectant solution for biological waste disposal.

5. Sterile forceps or tweezers.

6. Sterile scalpels and blades.

7. [³H]thymidine (2 μCi/mL; 1 Ci = 37 GBq) (Thermo Fisher Scientific, Chino, CA).

8. Kodak NTB-2 emulsion (Carestream Health, Rochester, NY).

9. Hematoxylin and eosin.

3   Methods

3.1   Establishment of Human Patient Tumors in Gelfoam® Histoculture

1. Immediately after surgery or biopsy, bring tumor sections in culture medium with Hanks' salts and fetal bovine serum to the laboratory.

2. Cut necrotic tissue away and mince the remaining healthy tumor tissue with scissors into 1–4 mm^3 fragments.

3. Place 5–10 tumor 1 mm^3 fragments on the Gelfoam® surface, where they tend to stick or become embedded in loose fiber structure.

4. Add cell culture medium until the upper part of the gel is reached but not covered.

5. Refeed the cultures twice a week.

3.2   Fixation, Embedding, and Staining of Tumors

1. Fix tumor tissue growing in the collagen gels in 10% formalin for at least 48 h.

2. Wash the fixed material for 1 h in slowly running tap water to rinse out the formalin and then process through a series of changes of ethanol (70–100%), with each change lasting 1 h.

3. Treat the material with xylene or chloroform.

4. Embed in paraffin.

5. After the paraffin is hardened, section the material at 5 μm and then dry for 10–15 min on a slide warmer.

6. The next day, stain the slides with Gill's hematoxylin no. 2 and eosin and then mount.

3.3   Autoradiography of Histocultured Tumors

1. Label the cells within the histocultured tumors metabolically, in the above-described medium, with [^{3}H]thymidine (1 μCi/ml, 60–90 Ci/mmol, ICN; 1 Ci = 37 GBq) for 24 h.

2. After labeling, remove the medium and wash the histoculture three times for 5 min each with nonradioactive medium.

3. Fix, dehydrate, embed, and section the histoculture in paraffin.

4. Deparaffinize the slides with xylene, coat with Kodak NTB-2 emulsion (diluted 1:1 with water) at 40 °C, and store in the dark at 40 °C for 1 week.

5. Develop the slides for 5 min in Kodak D-19, fix for 5 min in Rapid Fix, wash well in water, and then dry and stain.

6. Place approximately 10–15 sections of the ribbon on each side for autoradiographic processing and staining.

7. Determine replicating cells by the presence of bright green-reflecting silver grains over the cell nuclei that had incorporated [^{3}H]thymidine. Consider cells labeled if ten or more grains are present over the cell nucleus. Exclude stromal cells from the measurement by their morphological appearance.

8. When counting, scan each slide at ×40 or ×100 power to locate the areas of maximum label. Once these areas are located, determine the percentage of cells undergoing DNA synthesis in at least three of the visual fields for each tumor piece at ×200.

9. Calculate the growth fraction index (GFI) by dividing the number of labeled cancer cells by the number of total cancer cells.

4 Results

4.1 Long-Term Gelfoam® Histoculture of Human Patient Tumors

Sixty-five of 89 tumors explanted in Gelfoam® histoculture could be grown long term (*see* **Note 1**). For example, a patient melanoma was cultured on Gelfoam®. The macroscopic growth of this tumor is easily monitored because of the extent of melanin production (Fig. 1a).

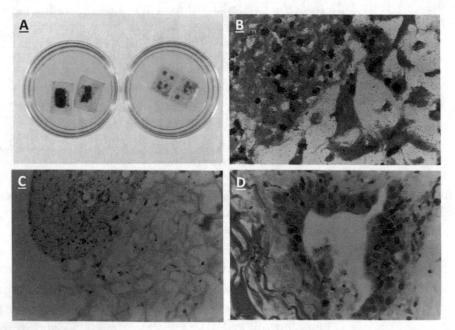

Fig. 1 (**a**) Human melanoma (patient 174) in Gelfoam® histoculture after 1 month in culture (left), compared to a typical start-up culture (right). Note that the melanoma has not only proliferated but also continued to produce melanin, indicating maintenance of differentiation in Gelfoam® histoculture. (**b**) In vitro incorporation of [³H]thymidine by Wilms tumor of the kidney (patient 181). After 6 days in culture, cells were labeled for 24 h with [³H]thymidine at 1 µCi/mL. Note the proliferating cells in the explant itself as well as proliferating cells within the Gelfoam® (autoradiogram, Giemsa counterstain; ×670). (**c**) Squamous cell carcinoma of the lung (patient 179), labeled with [³H]thymidine after 12 days in Gelfoam® histoculture. Note the high percentage of cells proliferating on the periphery of the tumor explants as well as cells proliferating within the Gelfoam® (autoradiogram, hematoxylin & eosin counterstain; ×170). (**d**) Colon cancer metastasis to the liver (patient 245). Labeling with [³H]thymidine was performed after 28 days in Gelfoam® histoculture. Note proliferating, densely packed cells and large, darkly staining nuclei (autoradiogram, hematoxylin & eosin; ×670) [8]

The measurement of proliferation of individual cells within human tumors in vitro is easily determined by autoradiography. An autoradiogram of a 6-day Gelfoam® histoculture from the Wilms tumor of a patient demonstrates that a large percentage of the cells incorporated thymidine during the 24-h labeling period (Fig. 1b). Another example is the 12-day Gelfoam® histoculture of a squamous-cell lung tumor of patient, where the majority of the proliferative cells are on the periphery of the tumor, while the interior contains a lesser number of proliferating cells, a situation similar to that in vivo (Fig. 1c). In both cases, not only cells within the explant itself but also cells that had migrated into the Gelfoam® were proliferative. Cellular proliferation in the 28-day Gelfoam® histoculture of a liver metastasis of a colon carcinoma is seen in Fig. 1d. Note the relatively high percentage of cells labeled and the large darkly staining nuclei in the densely packed cells, all indicating malignancy.

4.2   3-Dimensional Organization in Gelfoam® Histoculture

Gelfoam® histocultures of normal and malignant breast tissue from the same patient contain duct-like structures and more than one cell type. However, the cellular organization is diminished in the cancerous tissue, as can be seen from the cells surrounding the duct, which are no longer in parallel alignment (Fig. 2a, b). Thus, not only is general tissue organization maintained in Gelfoam® histoculture, but specific oncogenic changes are also maintained. Gelfoam®-histocultured normal and cancerous stomach tissue demonstrates that the cancerous tissue seems to have invaded the gel to a much greater degree than has the normal tissue. The nuclei of the cancerous tissue stain darker than the normal, which is frequently the case in vivo. In addition, the central areas of the explant seem to be necrotic (Fig. 2c, d) (*see* **Note 2**).

4.3   Multiple Cell Types from Individual Tumors In Vitro

Tumors are heterogeneic with regard to the cell types that they contain [10]. In Gelfoam® histoculture of individual growing tumors gives rise to distinctly different cell types, even to the extent that cells show seemingly different degrees of anchorage requirements. For instance, of the cells that migrated from the explant of the lung tumor of a patient, some grew in suspension in the medium (Fig. 3a), whereas others attached to the surface of the plastic culture dish (Fig. 3b). A highly malignant morphology of the lung cancer cells is consistent with the ability of the cells to grow in suspension. Cells from nine different tumors (three colon, one melanoma, one lung, one parotid adenoma, two lymph node metastases of unknown origin, and a lymph node metastasis of a thyroid papillary carcinoma), explanted as described above, were shown to also proliferate on the plastic surface of the culture dish as well as on and within Gelfoam®.

The proliferation kinetics of suspension-growing cells derived from Gelfoam® histoculture of a solid breast tumor and of a lymph

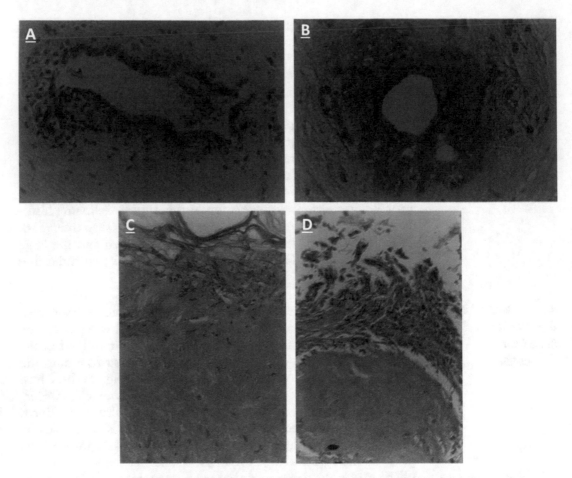

Fig. 2 (a) Normal breast tissue explant (patient 127) after 19 days in Gelfoam® histoculture. Note maintenance of cellular organization, in particular in the cells lining the duct (×500). (**b**) Breast tumor tissue explant (patient 127) after 7 days in Gelfoam® histoculture. Note the decrease in cellular organization, in particular in the cells lining the duct (×500). (**c**) Normal stomach tissue (patient 123) after 36 days in Gelfoam® histoculture (×340). (**d**) Cancerous stomach tissue (patient 123) after 36 days in Gelfoam® histoculture. Note the invasion of the tumor tissue into the Gelfoam® (×340) [8]

node metastasis of a rectal tumor were calculated. It can be seen that these solid tumor explants gave rise to cells that can proliferate continuously in suspension in the culture medium (Fig. 4).

4.4 Differentiated Functions of Tumor Expressed in Primary Gelfoam® Histoculture

An example of differentiated function maintenance growing in Gelfoam® histoculture is the production of melanin in the culture of melanoma (*see* **Note 3**). The heterogeneity in cell size and pigment production of a melanoma growing within the Gelfoam® after migration from the original explant is seen (Fig. 5a). Different cell types, growing as an attached culture, derived from the Gelfoam® histoculture of the omentum metastasis of the ovarian tumor of patient can also be seen (Fig. 5b).

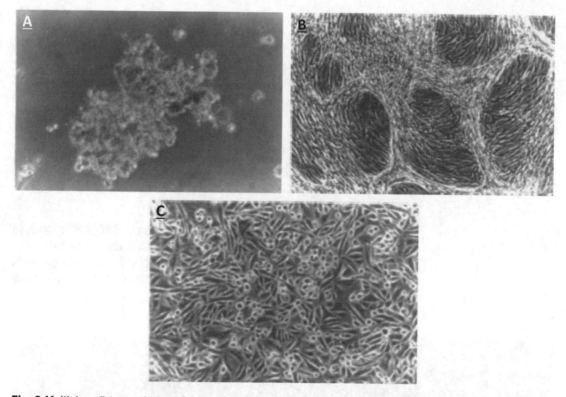

Fig. 3 Multiple cell types derived from lung carcinoma and growing in different phases. (**a**) Cells that have migrated from the original explant (patient 184) into the Gelfoam® to proliferate in suspension in the culture medium, after 21 days in Gelfoam® histoculture (×250). (**b**) Cells that have migrated from the same original explant (patient 184) to grow attached to the plastic surface of the culture dish, after 21 days in Gelfoam® histoculture (×70). (**c**) Cells from a different lung tumor explant (patient 237) that have migrated to grow attached to the plastic surface (×170) [8]

4.5 Tumorigenicity of Cancer Cells After Growth in Gelfoam® Histoculture

Five million melanoma cells derived from patient 174 and grown in Gelfoam® histoculture for 9 months were inoculated subcutaneously into three nude mice, each of which developed very large tumors within 2.5 months (Fig. 6) (*see* **Note 4**).

4.6 Summary of Results of Gelfoam® Histoculture of Human Tumors

Sixty-five patient tumors representing more than 17 types of human tumors were grown in Gelfoam® histoculture. The only major factor limiting culture seems to be bacterial infections, especially in the colon tumors. The tumors growing in Gelfoam® culture include those of the lung, colon, breast, bone, cervix, small bowel, rectum, testis, prostate, kidney, stomach, ovary, thyroid, and skin. Gelfoam® histoculture can be maintained for long periods, often for >100 days. Most important of all, the Gelfoam®-histocultured tumors in vitro maintain critical properties of the in vivo state [12, 13] (*see* **Note 5**).

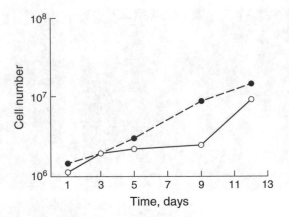

Fig. 4 Growth curves of cells derived from breast-cancer Gelfoam® histoculture (patient 187, *filled circle*) and from a lymph-node metastasis of an anal carcinoma in Gelfoam® histoculture (patient 130, *open circle*). Cells were grown in suspension in MEM containing 10% fetal bovine serum. Cells were enumerated in a Coulter counter [8]

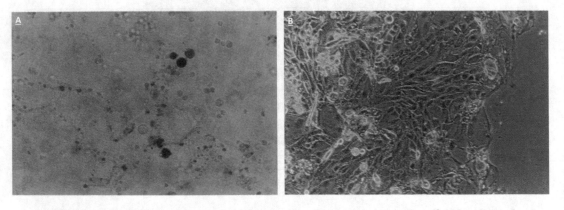

Fig. 5 (**a**) Melanoma cells (6 months in culture, from patient 174) growing in Gelfoam® histoculture after one transfer from the original Gelfoam® histoculture where the tumor was originally explanted. Note the heterogeneity in cell size and pigment production (×170). (**b**) Cells (84 days in culture) migrating from the explant on Gelfoam® of omentum metastasis of ovarian tumor (patient 227) to grow attached to the plastic surface of the cell culture dish. Note the apparent heterogeneity of the cells present (×170)

Please *see* Chapter 12 for cell cycle behavior of tumors in Gelfoam® histoculture.

5   Notes

1. The data presented here show that human tumors can grow at high frequency (64 of the 89 tumors explanted have grown) for prolonged periods in vitro.

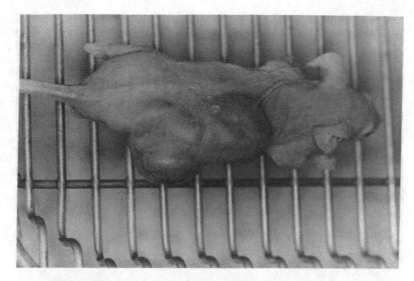

Fig. 6 Tumor formation in nude mice following inoculation of 5×10^6 melanoma cells of patient 174. The time from inoculation to the photograph was 85 days

2. Tumors explants maintain important in vivo properties, which include those that distinguish them from normal cells and tissues.

3. Melanoma cells from patient 174 produce large quantities of melanin in vitro, indicating long-term maintenance of differentiated function in primary culture [11].

4. Cells can be tumorigenic in nude mice after Gelfoam® histoculture.

5. The findings reviewed in the chapter here should be important for basic tumor biological studies as well as for clinical studies and applications.

References

1. Fraslin J, Kneip B, Vaulont S, Glaise D, Munnich A, Guguen-Guillouzo C (1985) Dependence of hepatocyte-specific gene expression on cell-cell interactions in primary culture. EMBO J 4:2487–2491

2. Clayton DF, Darnell JE Jr (1983) Changes in liver-specific compared to common gene transcription during primary culture of mouse hepatocytes. Mol Cell Biol 3:1552–1561

3. Lawler EM, Miller FR, Heppner GH (1983) Significance of three-dimensional growth patterns of mammary tissues in collagen gels. In Vitro 19:600–610

4. Leighton J, Justh G, Esper M, Kronenthal RL (1967) Collagen-coated cellulose sponge: three dimensional matrix for tissue culture of Walker tumor 256. Science 155:1259–1261

5. Yang J, Richards J, Bowman P, Guzman R, Enami J, McCormick K, Hamamoto S, Pitelka D, Nandi S (1979) Sustained growth and three-dimensional organization of primary mammary tumor epithelial cells embedded in collagen gels. Proc Natl Acad Sci U S A 76:3401–3405

6. Miller BE, Miller FR, Heppner GH (1985) Factors affecting growth and drug sensitivity of mouse mammary tumor lines in collagen gel cultures. Cancer Res 45:4200–4205

7. Miller BE, Miller FR, Heppner GH (1981) Interactions between tumor subpopulations affecting their sensitivity to the antineoplastic

agents cyclophosphamide and methotrexate. Cancer Res 41:4378–4381

8. Freeman AE, Hoffman RA (1986) In vivo-like growth of human tumors in vitro. Proc Natl Acad Sci U S A 83:2694–2698

9. Vescio RA, Redfern CA, Nelson TJ, Ugoretz S, Stern PH, Hoffman RM (1987) In vivo-like drug responses of human tumors growing in three-dimensional gel-supported primary culture. Proc Natl Acad Sci U S A 84:5029–5033

10. Bhatia S, Frangioni JV, Hoffman RM, Iafrate AJ, Polyak K (2012) The challenges posed by cancer heterogeneity. Nat Biotechnol 30:604–610

11. Lockshin A, Giovanella B, DeIpolyi PD, Williams LJ, Mendoza JT, Yim SO, Stehlin JS (1985) Exceptional lethality for nude mice of cells derived from a primary human melanoma. Cancer Res 45:345–350

12. Hoffman RM (2010) Histocultures and their use. In: Encyclopedia of life sciences. Wiley, Chichester. https://doi.org/10.1002/9780470015902.a0002573.pub2

13. Hoffman RM (2013) Tissue culture. In: Brenner's encyclopedia of genetics, 2nd edn, vol. 7. Elsevier, pp. 73–76.

Chapter 4

Expression and Targeting of Tumor Markers in Gelfoam® Histoculture: Potential Individualized Assays for Immuno-Oncology

Robert M. Hoffman and Fiorella Guadagni

Abstract

Tumor-specific antigens are important in the study of tumor biology, tumor diagnosis, and prognosis and as targets for tumor therapy. This chapter reviews patient colon, breast, and ovarian tumors in 3-dimensional Gelfoam® histoculture maintaining in vivo-like expression of the important tumor antigens, for example TAG-72 and CEA. We have also reviewed that fluorescent antibodies can target tumors in Gelfoam® histoculture, thereby providing an assay for individual patients for sensitivity to therapeutic antibodies which have become so important in immuno-oncology and other cancer therapies.

Key words Histoculture, Gelfoam, Tumor antigens, Expression, Monoclonal antibodies, Targeting, Immuno-oncology

1 Introduction

Previous observations showed that there is a very low expression of tumor antigens in monolayer cultures [1]. For example, the expression of the tumor antigen TAG-72 can be increased in vitro by culturing tumor cell lines as spheroids or in blocks of agar indicating that a 3D configuration of cells is necessary for antigen expression [2]. However, it is important to be able to culture tumor specimens with maintenance of tumor architecture and function for relatively long periods of time in order to study biological and clinical related properties of tumor antigens [3].

3D sponge-matrix histoculture first described by Leighton in 1951 [4] and further developed by us [5–27] enables human surgical cancer specimens to grow and be maintained with in vivo-like expression of TAG-72 and CEA for relatively long periods [3]. We have also demonstrated that prostate tissue on Gelfoam® histoculture also maintains expression of a key differentiation and diagnostic antigen, prostate-specific antigen (PSA) [15].

Robert M. Hoffman (ed.), *3D Sponge-Matrix Histoculture: Methods and Protocols*, Methods in Molecular Biology, vol. 1760, https://doi.org/10.1007/978-1-4939-7745-1_4, © Springer Science+Business Media, LLC, part of Springer Nature 2018

A major problem with clinical use of therapeutic monoclonal antibodies is that not only are tumors of the same histological type highly heterogeneous with regard to each other, but also intra-tumor heterogeneity exists with regard to many properties, including surface-associated antigen expression [1, 4]. Therefore, it is critical to develop systems in which to test the targeting of potentially diagnostic and therapeutic antibodies or their conjugates to human tumors, before clinical use. Although such tests could be carried out on frozen or paraffin-embedded sections of tumors, potentially more clinically-relevant information could be obtained by antibody-targeting studies on living tissues in Gelfoam® histo-culture [1, 4].

2   Materials

2.1   Culture Medium [6]

1. Culture medium (CM): RPMI 1640 with L-glutamine and phenol red, 100 μM modified Eagle's medium (MEM)-nonessential amino acids (Thermo Fisher Scientific), 1 mM sodium pyruvate, 50 μg/mL gentamycin sulfate, 2.5 μg/mL amphotericin B (Thermo Fisher Scientific).
2. 10% Fetal bovine serum (FBS)(GIBCO): We recommend testing different batches of FBS before purchasing a large stock.
3. 200 mM L-Glutamine (GIBCO).

2.2   Gelfoam® Sponge [6]

1. Gelfoam® Absorbable Collagen Sponge USP, 12–7 mm format (Pfizer, NDC: 0009-031508) as an example.

2.3   Materials for Preparation of Collagen Sponges [6]

1. Gelfoam® trimming aid: Double-edged blade.
2. Sterile Petri dish, 100 mm × 20 mm or wider.
3. Tissue transportation medium: Sterile phosphate buffer saline (PBS) or other solution with physiologic pH.
4. Sterile transportation container.
5. Sterile metal forceps or tweezers.
6. Sterile metal scissors.
7. Sterile flat weighing metal spatula.
8. 6-well culture plates.
9. Water-jacketed CO_2 incubator, set at 37 °C, 5% CO_2, ≥90% humidity.
10. Water bath.

2.4   Processing of Human Tumor Tissue [6]

1. Specimen of human tumors obtained from surgery.
2. Sterile Petri dish, 100 mm × 20 mm.
3. 70% ethanol solution.

4. Disinfectant solution for biological waste disposal.

5. Sterile forceps or tweezers.

6. Sterile scalpels and blades.

7. Hematoxylin and eosin.

2.5 Monoclonal Antibodies

1. Monoclonal antibodies B72.3 and COL-1 (Anti CEA).

2. Anti-prostate-specific antigen (PSA, Immunon, Pittsburgh, PA).

3. Anti-HLA-ABC (Miltenyi Biotec, San Diego, CA).

3 Methods

3.1 Gelfoam® Histoculture

1. Divide tumor into 1- to 2-mm-diameter pieces and place on top of Gelfoam® previously hydrated in cell culture medium.

2. Add culture medium to culture dishes such that the upper part of the Gelfoam® is not covered.

3.2 Immunohisto-chemical Studies [3, 12]

1. Embed tissues in paraffin using routine procedures. Cut 5 μm sections of each block and mount in gelatin-coated slides.

2. Stain sections with hematoxylin and eosin (H&E) to evaluate tumor type, grade, presence of necrosis, inflammation, and presence of mucin.

3. React slides with B72.3 and COL-1 or other monoclonal antibody at 40 μm/ml using a modification of the avidin-biotin-peroxidase complex technique using commercial reagents (Vector Laboratories, Burlingame, CA).

4. Use nonspecific IgC$_1$ as a negative control at a concentration of 40 μg/ml.

5. Determine peroxidase activity with freshly prepared diaminobenzidine (Sigma Co., St. Louis, MO) containing 0.1% hydrogen peroxide [3].

6. Evaluate each section for epithelial intracellular brown diaminobenzidine precipitate indicative of MAb binding. Assign the approximate percentage of positive staining carcinoma cells according to the number of carcinoma cells positive divided by the total number of cells present × 100 [3].

7. For PSA detection, block nonspecific reactivity with protein-blocking agent (PBA).

8. Pre-dilute antibody to prostate-specific antigen (PSA, Immunon, Pittsburgh, PA) (in 0.05 M TRIS in normal saline, pH 7.4 with a 0.2% sodium azide) and apply to sections of paraffin-sectioned or frozen-sectioned material.

9. Incubate the tissue for 30–60 min at 22 °C. After incubation, carry out standard immunostaining protocol [15].

10. For tumor targeting, use a fluorescent-dye-conjugated anti-CEA or other antibody [28–31].

11. For tumor targeting after 24 h of histoculture, add 4 µg/ml of fluorescein-labeled antibody for 1 h at 37 °C followed by five washes with culture medium. Analyze the antibody-treated histoculture with a scanning laser confocal microscope [12].

4 Results

4.1 Analysis of Tumor Antigen Expression in Gelfoam® Histoculture

Thirteen-day histocultures of a patient colon tumor were fixed and prepared for immunohistochemical analysis as described above. The human patient colon carcinoma was immunohistochemically stained for expression of TAG-72 and CEA. The histocultured tissue demonstrated that high expression of TAG-72 can be maintained for a relatively long period in Gelfoam® histoculture (Fig. 1a). This is contrary to studies in other laboratories on TAG-72 expression in monolayer cultures of tumor cell lines which indicate lack of expression of TAG-72 which can be reactivated by growing the cells in vivo or partially reactivated by growing the cells in vitro as spheroids or in soft agar [3].

A 13-day histocultured patient colon carcinoma from another patient also had a high reactivity with anti-CEA antibody COL-1 (Fig. 1b) indicting high maintenance of CEA expression in histoculture of colon carcinomas (*see* **Note 1**). These results, along with the TAG-72 results, indicate the generality of high levels of tumor antigen expression in histoculture of colon carcinomas [3].

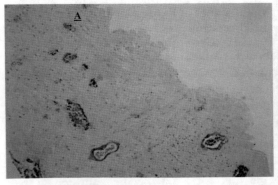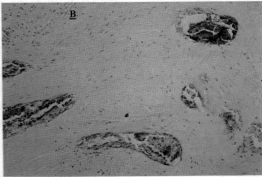

Fig. 1 Immunohistochemical analysis of human colon carcinoma from patient 165. (**a**) 13-day histocultures of colon cancer stained with 40 µg/ml B72.3 monoclonal antibody specific for TAG-72. (**b**) 13-day histoculture of colon cancer stained with 40 µg/ml COL-1 monoclonal antibody specific for CEA. The brown stain indicates the presence of the tumor antigen in each case [3]

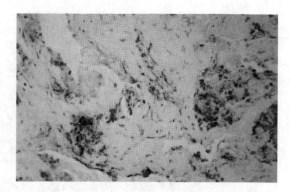

Fig. 2 Immunohistochemical analysis of the human ovarian cancer from patient 211 after 13-day histoculture using B72.3 monoclonal antibody at 40 µg/mL. The brown stain indicates the presence of the tumor antigen [3]

4.2 Analysis of Tumor Antigen Expression in Ovarian Cancer in Gelfoam® Histoculture

An ovarian tumor that was histocultured for 13 days expressed high levels of TAG-72 as shown by immunohistochemical staining using the B72.3 monoclonal antibody (Fig. 2) [3] (*see* **Note 2**).

These results indicate that the tumor-specific antigens such as TAG-72 and CEA can be expressed at relatively high levels for periods of at least 13 days in 3D Gelfoam histoculture of colon cancer. In addition, TAG-72 can be expressed at high levels for a relatively long period in 3D Gelfoam® histoculture of an ovarian tumor. An additional two colon tumors, two breast tumors, and three ovarian tumors in 13–21-day Gelfoam® histocultures also expressed TAG-72 and CEA in a similar manner as precultured material (data not shown). Additional data not shown demonstrated that TAB-72 is still expressed after 4 weeks of histoculture [3] (*see* **Note 3**).

4.3 Analysis of Tumor Antigen Expression in Benign Prostate Hyperplasia (BPH) IN Gelfoam® Histoculture

Prostate-specific antigen (PSA) was highly expressed in histocultured BPH specimens. Thus, an important differentiation antigen of the prostate can be continually expressed in Gelfoam histoculture (Fig. 3) [15].

4.4 Targeting of Tumor Antigens in Gelfoam Histoculture

Anti-HLA-ABC MAb targeted a human patient ovarian adenocarcinoma in Gelfoam histoculture (Fig. 4). There was extensive targeting of the antibody to the human tumor. The antibody, however, did not bind the tumor tissues homogeneously, and some areas were stained more intensively than others. The glandular nature of the histocultured tumor tissue can be seen in three dimensions by confocal microscopy demonstrating antibody bound throughout most of the tissue. A nonspecific isotype-matched antibody resulted in only negligible staining, thereby confirming the specificity of the binding of the anti-HLA-ABC MAb [12] (Fig. 4).

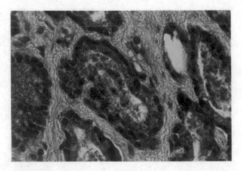

Fig. 3 Prostate-specific antigen (PSA) expression in sections of histocultured prostate tissue from a patient with benign prostate hypertrophy (BPH). Tissue was histocultured for 5 days and stained with antibodies to PSA as described in the text. The red-brown cytoplasmic staining indicates expression of PSA in the cells of the histocultured prostate [15]

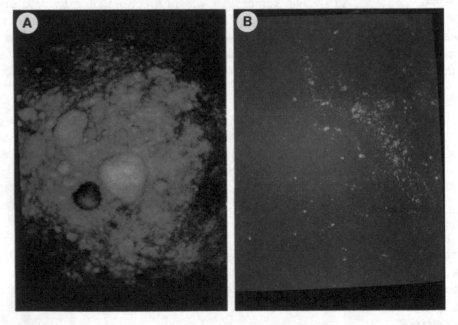

Fig. 4 Antibody targeting of a patient ovarian cancer in Gelfoam® histocutlure. Anti-HLA-ABC MAb was added at 4 μg/ml to Gelfoam®-supported histocuhure. After 1-h incubation, the tissue was extensively washed and analyzed with a BioRad MRC 600 scanning laser confocal microscope with an argon laser (0.4). (**b**) demonstrates the results when a nonspecific isotype-matched antibody was used in place of the anti-HLA-ABC MAb [12]

This chapter reviews that tumor antigen expression can be maintained for long periods in patient cancer tissue in Gelfoam® histoculture. The chapter also suggests that tumor antigens can be targeted by labeled monoclonal antibodies in Gelfoam® histoculture which can provide an individual patient assay for potential efficacy of fluorescence-guided surgery [28–31] and for response to immuno-oncology drugs [32–36]. So-called "immune-check-point inhibitors" have been effective for several cancers by stimulating anti-tumor immune responses [34, 35]. Ipilimumab, an anticytotoxic T-lymphocyte-associated antigen 4 (CTLA-4) antibody, and nivolumab, an anti-programmed death-1 (PD-1) antibody, have improved survival in patients with melanoma and early results suggest that their combination further enhances antitumor activity and survival [34, 36]. Their evaluation in Gelfoam® culture could prove useful.

5   Notes

1. COL-1 reacts with CEA, a 180,000-molecular-weight glycoprotein which is present on a high percentage of carcinomas of the gastrointestinal tract [3, 37–39].

2. B72.3 reacts with the tumor-associated glycoprotein 72 (TAG-72), a high-molecular-weight glycoprotein which is found in a high percentage of colorectal, stomach, breast, ovarian, and lung carcinomas [40].

3. TAG-72 is rarely expressed in normal tissue with the exception of secretary endometrium [41] and transitional colonic mucosa [42]. No TAG-72 has been found during the proliferative or resting phases of the endometrium.

References

1. Colcher D, Horand-Hand P, Nuti M, Schlom J (1981) A spectrum of monoclonal antibodies reactive with human mannary tumor cells. Proc Natl Acad Sci 73:3199

2. Horand-Hand P, Clocher D, Salomon D, Ridge J, Noguchi P, Schlom J (1985) Influence of spatial configuration of carcinoma cell populations on the expression of a tumor-associated glycoprotein. Cancer Res 45:833–840

3. Guadagni F, Roselli M, Hoffman RM (1991) Maintenance of expression of tumor antigens in three-dimensional *in vitro* human tumor gel-supported histoculture. Anticancer Res 11:543–546

4. Guadagni F, Roselli M, Amato T, Cosimelli M, Mannella E, Perri P, Abbolito MR, Cavaliere R, Colcher D, Greiner JW, Schlom J (1991) Tumor-associated glycoprotein-72 serum levels complement carcinoembryonic antigen levels in monitoring patients with gastrointestinal carcinoma. A longitudinal study. Cancer 68:2443–2450

5. Freeman A, Hoffman RM (1986) *In vivo*-like growth of human tumors *in vitro*. Proc Natl Acad Sci U S A 83:2694–2698

6. Vescio RA, Redfern CH, Nelson TJ, Ugoretz S, Stern PH, Hoffman RM (1987) *In vivo*-like drug response of human tumors growing in three-dimensional, gel-supported, primary culture. Proc Natl Acad Sci U S A 84:5029–5033

7. Hoffman RM, Monosov AZ, Connors KM, Herrera H, Price JH (1989) A general native-state method for determination of proliferation capacity of human normal and tumor tissues *in vitro*. Proc Natl Acad Sci U S A 86:2013–2017

8. Vescio RA, Connors KM, Youngkin T, Bordin GM, Robb JA, Umbreit JN, Hoffman RM (1990) Cancer biology for individualized cancer therapy: Correlation of growth fraction index in native-state culture with tumor grade and stage. Proc Natl Acad Sci U S A 87:691–695

9. Vescio RA, Connors KM, Bordin GM, Robb JA, Youngkin T, Umbreit JN, Hoffman RM

(1990) The distinction of small cell and non-small cell cancer by growth in native-state histoculture. Cancer Res 50:6095–6099

10. Hoffman RM (1991) Three-dimensional histoculture: origins and applications in cancer research. Cancer Cells 3:86–92

11. Vescio RA, Connors KM, Kubota T, Hoffman RM (1991) Correlation of histology and drug response of human tumors grown in native-state three-dimensional histoculture and in nude mice. Proc Natl Acad Sci U S A 88:5163–5166

12. Guadagni F, Li L, Hoffman RM (1992) Targeting antibodies to live tumor tissue in 3-D histoculture. In Vitro Cell Dev Biol 28A:297–299

13. Geller J, Sionit LR, Connors KM, Hoffman RM (1992) Measurement of androgen sensitivity in the human prostate in *in vitro* three-dimensional histoculture. Prostate 21:269–278

14. Hoffman RM (1993) To do tissue culture in two or three dimensions? That is the question. Stem Cells 11:105–111

15. Geller J, Sionit LR, Connors KM, Youngkin T, Hoffman RM (1993) Expression of prostate-specific antigen in human prostate specimens in *in vitro* three dimensional histoculture. In Vitro Cell Dev Biol 29A:523–524

16. Furukawa T, Kubota T, Hoffman RM (1995) Clinical applications of the histoculture drug response assay. Clin Cancer Res 1:305–311

17. Singh B, Li R, Xu L, Poluri A, Patel S, Shaha AR, Pfister D, Sherman E, Hoffman RM, Shah J (2002) Prediction of survival in patients with head and neck cancer using the histoculture drug response assay. Head Neck 24:437–442

18. Flowers JL, Hoffman RM, Driscoll TA, Wall ME, Wani MC, Manikumar G, Friedman HS, Dewhirst M, Colvin OM, Adams DJ (2003) The activity of camptothecin analogues is enhanced in histocultures of human tumors and human tumor xenografts by modulation of extracellular pH. Cancer Chemother Pharmacol 52:253–261

19. Jung PS, Kim DY, Kim MB, Lee SW, Kim JH, Kim YM, Kim YT, Hoffman RM, Nam JH (2013) Progression-free survival is accurately predicted in patients treated with chemotherapy for epithelial ovarian cancer by the histoculture drug response assay in a prospective correlative clinical trial at a single institution. Anticancer Res 33:1029–1034

20. Mii S, Duong J, Tome Y, Uchugonova A, Liu F, Amoh Y, Saito N, Katsuoka K, Hoffman RM (2013) The role of hair follicle nestin-expressing stem cells during whisker sensory-nerve growth in long-term 3D culture. J Cell Biochem 114:1674–1684

21. Mii S, Uehara F, Yano S, Tran B, Miwa S, Hiroshima Y, Amoh Y, Katsuoka K, Hoffman RM (2013) Nestin-expressing stem cells promote nerve growth in long-term 3-dimensional Gelfoam®-supported histoculture. PLoS One 8:e67153

22. Yano S, Miwa S, Mii S, Hiroshima Y, Uehara F, Yamamoto M, Kishimoto H, Tazawa H, Bouvet M, Fujiwara T, Hoffman RM (2014) Invading cancer cells are predominantly in G_0/G_1 resulting in chemoresistance demonstrated by real-time FUCCI imaging. Cell Cycle 13:953–960

23. Mii S, Amoh Y, Katsuoka K, Hoffman RM (2014) Comparison of nestin-expressing multipotent stem cells in the tongue fungiform papilla and vibrissa hair follicle. J Cell Biochem 115:1070–1076

24. Tome Y, Uehara F, Mii S, Yano S, Zhang L, Sugimoto N, Maehara H, Bouvet M, Tsuchiya H, Kanaya F, Hoffman RM (2014) 3-dimensional tissue is formed from cancer cells in vitro on Gelfoam®, but not on Matrigel™. J Cell Biochem 115:1362–1367

25. Kim KY, Chung BW, Yang I, Kim MB, Hoffman RM (2014) Independence of cytotoxic drug sensitivity profiles and receptor subtype of invasive ductal breast carcinoma demonstrated by the histoculture drug response assay (HDRA). Anticancer Res 34:7197–7202

26. Yano S, Miwa S, Mii S, Hiroshima Y, Uehara F, Kishimoto H, Tazawa H, Zhao M, Bouvet M, Fujiwara T, Hoffman RM (2015) Cancer cells mimic *in vivo* spatial-temporal cell-cycle phase distribution and chemosensitivity in 3-dimensional Gelfoam® histoculture but not 2-dimensional culture as visualized with real-time FUCCI imaging. Cell Cycle 14:808–819

27. Yano S, Takehara K, Miwa S, Kishimoto H, Tazawa H, Urata Y, Kagawa S, Bouvet M, Fujiwara T, Hoffman RM (2017) GFP labeling kinetics of triple-negative human breast cancer by a killer-reporter adenovirus in 3D Gelfoam® histoculture. In Vitro Cell Dev Biol Anim 53:479–482

28. Metildi CA, Tang CM, Kaushal S, Leonard SY, Magistri P, Tran Cao HS, Hoffman RM, Bouvet M, Sicklick JS (2013) In vivo fluorescence imaging of gastrointestinal stromal tumors using fluorophore-conjugated anti-KIT antibody. Ann Surg Oncol 20(Suppl 3):693–700

29. Metildi CA, Kaushal S, Pu M, Messer KA, Luiken GA, Moossa AR, Hoffman RM, Bouvet M (2014) Fluorescence-guided surgery with a fluorophore-conjugated antibody to carcinoembryonic antigen (CEA), that highlights the tumor, improves surgical resection and

increases survival in orthotopic mouse models of human pancreatic cancer. Ann Surg Oncol 21:1405–1411

30. Metildi CA, Kaushal S, Luiken GA, Talamini MA, Hoffman RM, Bouvet M (2014) Fluorescently-labeled chimeric anti-CEA antibody improves detection and resection of human colon cancer in a patient-derived orthotopic xenograft (PDOX) nude mouse model. J Surg Oncol 109:451–458

31. Metildi CA, Kaushal S, Luiken GA, Hoffman RM, Bouvet M (2014) Advantages of fluorescence-guided laparoscopic surgery of pancreatic cancer labeled with fluorescent anti-carcinoembryonic antigen antibodies in an orthotopic mouse model. J Am Coll Surg 219:132–141

32. Weber J, Mandala M, Del Vecchio M, Gogas HJ, Arance AM, Cowey CL, Dalle S, Schenker M, Chiarion-Sileni V, Marquez-Rodas I, Grob JJ, Butler MO, Middleton MR, Maio M, Atkinson V, Queirolo P, Gonzalez R, Kudchadkar RR, Smylie M, Meyer N, Mortier L, Atkins MB, Long GV, Bhatia S, Lebbé C, Rutkowski P, Yokota K, Yamazaki N, Kim TM, de Pril V, Sabater J, Qureshi A, Larkin J, Ascierto PA; CheckMate 238 Collaborators (2017) Adjuvant Nivolumab versus Ipilimumab in Resected Stage III or IV Melanoma. N Engl J Med 377:1824–1835

33. Wolchok JD, Chiarion-Sileni V, Gonzalez R, Rutkowski P, Grob JJ, Cowey CL, Lao CD, Wagstaff J, Schadendorf D, Ferrucci PF, Smylie M, Dummer R, Hill A, Hogg D, Haanen J, Carlino MS, Bechter O, Maio M, Marquez-Rodas I, Guidoboni M, McArthur G, Lebbé C, Ascierto PA, Long GV, Cebon J, Sosman J, Postow MA, Callahan MK, Walker D, Rollin L, Bhore R, Hodi FS, Larkin J (2017) Overall survival with combined nivolumab and ipilimumab in advanced melanoma. N Engl J Med 377:1345–1356

34. Johnson DB, Balko JM, Compton ML, Chalkias S, Gorham J, Xu Y, Hicks M, Puzanov I, Alexander MR, Bloomer TL, Becker JR, Slosky DA, Phillips EJ, Pilkinton MA, Craig-Owens L, Kola N, Plautz G, Reshef DS, Deutsch JS, Deering RP, Olenchock BA, Lichtman AH, Roden DM, Seidman CE, Koralnik IJ, Seidman JG, Hoffman RD, Taube JM, Diaz LA Jr, Anders RA, Sosman JA, Moslehi JJ (2016) Fulminant myocarditis with combination immune checkpoint blockade. N Engl J Med 375:1749–1755

35. Wolchok JD (2015) PD-1 blockers. Cell 162:937

36. Hodi FS, O'Day SJ, McDermott DF, Weber RW, Sosman JA, Haanen JB, Gonzalez R, Robert C, Schadendorf D, Hassel JC, Akerley W, van den Eertwegh AJ, Lutzky J, Lorigan P, Vaubel JM, Linette GP, Hogg D, Ottensmeier CH, Lebbé C, Peschel C, Quirt I, Clark JI, Wolchok JD, Weber JS, Tian J, Yellin MJ, Nichol GM, Hoos A, Urba WJ (2010) Improved survival with ipilimumab in patients with metastatic melanoma. N Engl J Med 363:711–723

37. Sikorska H, Shuster J, Gold P (1988) Clinical applications of carcinoembryonic antigen. Cancer Detect Prev 12:321–355

38. Muraro R, Wunderlich D, Thor A, Lundy J, Noguchi P, Cunningham R, Schlom J (1985) Definition by monoclonal antibodies of a repertoire of epitopes on carcinoembryonic antigen differentially expressed in human colon carcinomas versus normal adult tissues. Cancer Res 45:5769–5780

39. Kuroki M, Greiner JW, Simpson JF, Primus FJ, Guadagni F, Schlom J (1989) Serologic mapping and biochemical characterization of the carcinoembryonic antigen epitopes using fourteen distinct monoclonal antibodies. Int J Cancer 44:208–218

40. Thor A, Ohuchi N, Szpak CA, Johnston WW, Schlom J (1986) Distribution of oncofetal antigen tumor-associated glycoprotein-72 defined by monoclonal antibody B72.3. Cancer Res 46:3118–3124

41. Thor A, Viglione MJ, Muraro R, Ohuchi N, Schlom J, Gorstein F (1987) Monoclonal antibody B72.3 reactivity with human endometrium: a study of normal and malignant tissues. Int J Gynecol Pathol 6:235–247

42. Wolf BC, D'Emilia JC, Salem RR, DeCoste D, Sears HF, Gottlieb LS, Steele GD Jr (1989) Detection of the tumor-associated glycoprotein antigen (TAG-72) in premalignant lesions of the colon. J Natl Cancer Inst 81:1913–1917

Chapter 5

Development of the Histoculture Drug Response Assay (HDRA)

Robert M. Hoffman and Robert A. Vescio

Abstract

The histoculture drug response assay (HDRA) was developed using Gelfoam® histoculture of all tumor types. Twenty tumor classes, including all the major ones, have been histocultured on Gelfoam® and tested for drug response. Quantitative and qualitative results show increasing cell kill with rising cytotoxic drug concentration, differential drug sensitivities of multiple cell types within individual cultured tumors, differential sensitivities to a single drug of a series of tumors of the same histopathological classification, differential sensitivities of individual tumors to a series of drugs, and sensitivity patterns of various tumor types similar to the sensitivities found in vivo. Therefore, the results indicated that precise therapeutic data can be obtained from tumor specimens growing in Gelfoam® histoculture in vitro for the individual cancer patient as well as for rational and relevant screening for novel agents active against human solid tumors.

Key words 3D Gelfoam® culture, Patient tumors, Drug response, Personalized therapy, Precision oncology

1 Introduction

"Histologically identical tumors often differ in their response to treatment"

Black and Speer, 1954 (JNCI 14, 1147–1158)

A major clinical problem is that cancers that are classified as identical according to their histopathological characteristics are nonetheless highly individual in their drug sensitivities making it difficult to individualize treatment [1, 2]. To overcome the problem, many attempts have been made to develop in vitro drug sensitivity tests for individual cancer patients about to undergo chemotherapy and to screen for new anticancer agents. Plating of dissociated tumor cells in soft agar [3–9] and monolayer cultures [10–12] does not represent in vivo conditions and may preclude cell types present in the original tumor from growing. Multicellular spheroids that are three-dimensional have been used for drug sensitivity testing [13, 14], but these also involve dissociation of cells

Robert M. Hoffman (ed.), *3D Sponge-Matrix Histoculture: Methods and Protocols*, Methods in Molecular Biology, vol. 1760, https://doi.org/10.1007/978-1-4939-7745-1_5, © Springer Science+Business Media, LLC, part of Springer Nature 2018

from the tumor and reassociation into structures that do not resemble the original histology of the tumor. Short-term in vitro assays of drugs on non-cultured non-dissociated tumor specimens are not physiological [15, 16]. Cells dissociated from solid tumors are also not under physiological conditions [17].

Leighton in 1951 brought tissue culture closer to reflecting the in vivo situation by introducing a three-dimensional sponge matrix system. Leighton was able to grow tumors with maintenance of in vivo-like architecture [18–32] (please *see* Chapter 1 of this volume).

Browning and Trier [33] and Schiff [34] also used 3D culture conditions and kept cultured tissues above the culture medium by explanting the tissue on a matrix. Autrup [35] used a 95% O_2/5% CO_2 as an atmosphere to enhance oxygenation and used gelatin sponges as a growth support [36].

Sherwin and Richters [37] have termed the approach of maintaining in vivo-like tissue architecture in vitro as "histoculture." Their approach was to explant 2 mm^2 tissues about 1 mm thick, which would allow for optimal diffusion [38], which is very important as pointed out in the original tissue culture article by Alexis Carrel [39]. These ideas were further developed by Acedia et al. [40, 41] who cultured embryonic lung tissue on sponges derived from pigskin. This approach was also further developed by Douglas et al. [42].

We have developed three-dimensional Gelfoam® histoculture that is general and grows most human tumors obtained directly from surgery or biopsy. The culture system meets important criteria of in vivo growth, including maintenance of tissue structure [19]. This chapter reviews the original development of the histoculture drug-response assay (HDRA) based on Gelfoam® 3D histoculture of tumors [43].

2   Materials [43]

2.1   Culture Medium

1. Culture medium (CM): RPMI1640 with L-glutamine and phenol red, 100 µM modified Eagle's medium (MEM) nonessential amino acids (Thermo Fisher Scientific), 1 mM sodium pyruvate, 50 µg/ml gentamycin sulfate, 2.5 µg/ml amphotericin B (Thermo Fisher Scientific).

2. 10% Fetal bovine serum (FBS) (GIBCO): We recommend testing different batches of FBS before purchasing a large stock.

3. 200 mM L-Glutamine (GIBCO).

2.2   Gelfoam® Sponge

1. Gelfoam® sources: Gelfoam sponges comprised of absorbable gelatin prepared from purified porcine skin (Pfizer, NDC: 0009-031508) (*see* **Note 1**).

2.3 Materials for Preparation of Collagen Sponges

1. Gelfoam® trimming aid: Double-edged blade.
2. Sterile Petri dish, 100 mm × 20 mm or wider.
3. Tissue transportation medium: Sterile phosphate buffer saline (PBS) or other solution with physiologic pH.
4. Sterile transportation container.
5. Sterile metal forceps or tweezers.
6. Sterile metal scissors.
7. Sterile flat weighing metal spatula.
8. 6-well culture plates.
9. Water-jacketed CO_2 incubator, set at 37 °C, 5% CO_2, ≥90% humidity.
10. Water bath.

2.4 Processing of Human Tumor Tissue

1. Specimen of human tumors obtained from surgery.
2. Sterile Petri dish, 100 mm × 20 mm.
3. 70% ethanol solution.
4. Disinfectant solution for biological waste disposal.
5. Sterile forceps or tweezers.
6. Sterile scalpels and blades.
7. [³H]thymidine (2 µCi/mL; 1 Ci = 37 GBq).
8. NTB-2 emulsion (Kodak).
9. Hematoxylin and eosin.

3 Methods [43]

1. Mince surgically removed patient tumors into 2–3 mm³ pieces in previously hydrated Gelfoam®.
2. Add culture medium to dishes or wells such that the upper part of the Gelfoam® is not covered.
3. Use drugs with the therapeutic concentrations and exposure times listed in Table 1 [44, 45].
4. After a period of 3 days to recover from any transient effects of the drugs, measure cell proliferation by administering [³H]thymidine (2 µCi/mL; 1 Ci = 37 GBq) to the culture [46] for 4 days. Cellular DNA is labeled in any cells undergoing replication within the tumors.
5. After 4 days of labeling, wash the cultures with phosphate-buffered saline, place in histology capsules, and fix in 10% formalin.
6. Dehydrate the cultures, embed in paraffin, and section by standard methodology.

Table 1
Drug concentrations and exposure times used in vitro [43]

Drug	Therapeutic (1×) concentration	Exposure time, h
Doxorubicin (DOX)[a]	29 ng/mL	24
BCNU	0.2 µg/mL	3–24
Cisplatinum (CDDP)	1.5 µg/mL	24
Melphalan (MEL)	0.5 µg/mL	5–24
Methotrexate (MTX)	2.25 µg/mL	24
Mitomycin C	100 µg/mL	1.5–24
5-Fluorouracil (5-FU)	4.0 µg/mL	1–24
Vincristine	23 ng/mL	2.5–24
VP-16	4.8 µg/mL	24
Interferon (IFN)	3×10^3 units/mL	24
Deoxycoformycin (dCOF)	26.8 µg/mL	5–24

The drug concentrations and exposure times were calculated from pharmacological data to simulate in vitro the drug concentrations achieved in vivo. Drug exposure times are based on the plasma half-life of the chemotherapeutic agents [44, 45]. Drugs were given after tumors were cultured for at least 4 days
[a]Generic name = doxorubicin hydrochloride [43]

7. After the slides are deparaffinized, prepare for autoradiography by coating with Kodak NTB-2 emulsion in the dark, expose for 5 days, and then develop the slides.

8. After rinsing, stain the slides with hematoxylin and eosin.

9. Analyze the slides by determining the percentage of cells undergoing DNA synthesis in treated vs. untreated tumor cultures. Replicating cells are identified by the presence of black grains over their nuclei due to exposure of the NTB-2 emulsion to radioactive DNA.

4 Results

4.1 Differential Effects of Various Drugs on a Patient Ovarian Cancer in Gelfoam® Histoculture (See Note 1)

The responses to seven different drugs of the cultured ovarian cystadenoma of patient 414 are shown in Fig. 1. There are two populations of cells present. The large-nuclei-containing typical ovarian carcinoma cells [47] were actively dividing, as evidenced by the black grains over them; the small cells were not dividing. The efficacy of various drugs was determined at three different concentrations including those that correspond to usual clinically-achievable plasma concentrations. By comparing the number of proliferating

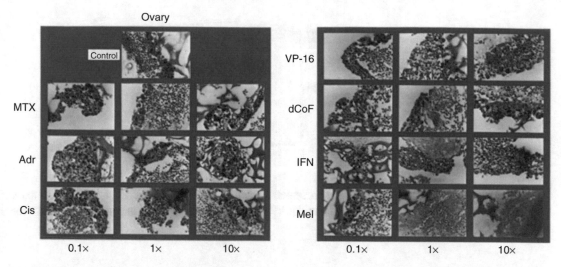

Fig. 1 Drug responses of cultures of ovarian cancer of patient 414. Autoradiograms were prepared from slides of cultures treated with the indicated drugs and labeled with [3^H]thymidine. Autoradiograms were counter-stained with hematoxylin and eosin. Black grains over purple-stained nuclei indicate the uptake of radioactivity into the cells and DNA synthesis. Note the resistance of cultured tumor to MTX (1 × concentration, 2.25 μg/mL; 24-h exposure), Adr (1 × concentration, 29 ng/mL; 24-h exposure), Cis (1 × concentration, 1.5 μg/mL; 24-h exposure), VP-16 (I × concentration, 4.8 μg/mL; 24-h exposure), dCof (I × concentration, 26.8 μg/mL; 24-h exposure), and IFN (I × concentration, 3 × 10^3 units/mL; 24-h exposure), and the sensitivity to Mel (I × concentration, 0.5 μg/mL; 5-h exposure) (×173) [43]

cells in the drug-treated cultures to the control, it can be determined which of these drugs is effective for this tumor. The resistance to doxorubicin (DOX) and cisplatinum (CDDP) is not surprising in light of the patient's previous treatment failure with these drugs. However, as shown in the bottom row, the drug melphalan (MEL) seems highly effective for the tumor (Fig. 2) with the 1× and 10× concentrations eliminating cellular proliferation and most of the cells themselves, respectively [43]. For the ovarian carcinoma, only MEL was an effective drug of the seven that were tried.

4.2 Efficacy of Drugs on a Patient Breast Cancer Gelfoam® Histoculture (See Note 2)

The radioactively labeled nuclei indicated that there was a large population of cells in the tumor that were resistant to DOX. However, the 1× concentration of methotrexate (MTX) and cisplatinum (CDDP) were effective in reducing cell proliferation within the cultured tumor [43].

A histocultured colon cancer liver metastasis was resistant to very high concentrations of DOX and 5-fluorouacil (5-FU) (Fig. 3). The autoradiogram indicates resistance to the 10× concentration of 5-FU. Thus it is possible to determine the proliferative inhibition by drugs on individual cells within the tissue structure of the cultured tumors [43].

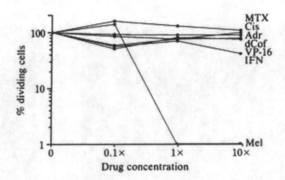

Fig. 2 Graphic representation of drug responses of cultures of ovarian carcinoma of patient 414 (*see* Fig. 1 for autoradiographic data of in vitro drug response for this patient) [43]

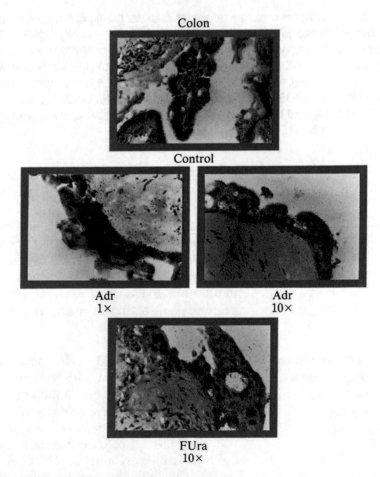

Fig. 3 Drug responses of cultures of a colon carcinoma liver metastasis of patient 337. *See* Fig. 1 for details. Note the resistance of the cultured tumor to high levels of Adr and FUra (×400) [43]

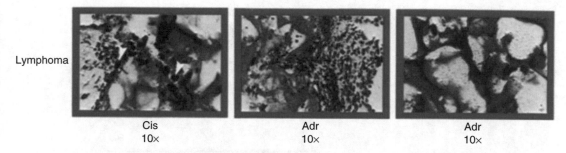

Lymphoma

Cis
10×

Adr
10×

Adr
10×

Fig. 4 Drug responses of different cell types of lymphoma of patient 277. *See* text for details. Note that cells within the tissue structure and those migrating into the Gelfoam® are both resistant to Cis (*see* arrows). However, only the cells migrating into the Gelfoam® are resistant to Adr. In the autoradiogram in the center, the nuclei within the tissue structure itself are darkly stained but not autoradiographically labeled and thereby responding to Adr as opposed to some of the nuclei of cells migrating into the Gelfoam®, which are labeled autoradiographically and resistant to Adr (×400) [43]

4.3 Demonstration of Differential Efficacy of a Single Drug on Various Cell Types Within a Single Tumor (See Note 3)

An axillary lymph node involved with lymphoma was histocultured on Gelfoam® and treated with the 1× concentration of CDDP. There was a large degree of proliferation of the cells in the tumor explant tissue structure itself as well as in cells that invaded the supporting Gelfoam® matrix, indicating resistance of both cell types to CDDP (Fig. 4). An ultra-high concentration of DOX caused cessation of proliferation of cells in the tumor explant. However, DOX, even at ten times the therapeutic level, did not inhibit the proliferation of cells that have invaded the Gelfoam® [43].

A colon cancer was histocultured on Gelfoam®. The cells within the explant tumor-tissue structure were resistant to DOX as well as to 5-FU even at the ultra-high concentrations, while cells migrating into the Gelfoam® were sensitive to these drugs [43].

These data indicate that the cells that migrate into the Gelfoam® may be of a different quality from those that do not, and their differences include drug sensitivity [43].

4.4 Demonstration of Differential Sensitivity of Members of a Set of Tumors of a Single Histopathological Type to an Individual Drug (See Note 4)

A series of breast cancer of the same histopathological type from different individuals was histocultured on Gelfoam®. The percentage of dividing cells relative to the control was plotted as a function of DOX concentration for the series of cultured breast tumors from different patients. The range of sensitivity was very large, which reflects the clinical situation for breast cancer patients in general (Fig. 5) (*see* **Note 4**) [43].

5 Notes

1. The drug sensitivities of the tumors thus far tested in vitro also resemble the overall clinical pattern [48], which indicates the in vitro drug sensitivities of 15 different types of cancers. Later

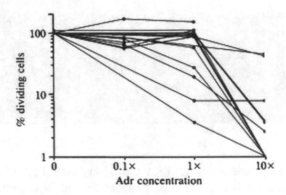

Fig. 5 Graphic of responses of a series of cultured breast tumors to Adr (*see* text for details) [43]

clinical correlation studies demonstrated the precision of the HDRA to identify both effective and ineffective drugs for the patient (*see* Chaps. 7–9 of the present volume).

2. It is important to note that the HDRA described here is a general one and allows drug-response data to be obtained in all types of solid tumors at high frequency. The fact that tumors can be cultured for long periods [49] allows long-term testing, and cells that still retain proliferative capacity after drug treatment can be detected.

3. Given the heterogeneous nature of individual tumors [50, 51], it is important to be able to measure the drug sensitivities of specific cell types of cultured tumors, as shown here. This type of analysis is made possible by the fact that Gelfoam® histoculture maintains three-dimensional tissue structure and apparantly all cell types of the tumor in vivo and can be observed histologically in autoradiograms.

4. Tumors classified into a single histopathological type can have vastly different drug responses and therefore, precision therapy, such as that enabled by the HDRA, is necessary for patients [52].

References

1. Chabner BA (ed) (1983) Rational basis for chemotherapy. Liss, New York
2. Finlay GJ, Baguley BC (1984) The use of human cancer cell lines as a primary screening system for antineoplastic compounds. Eur J Cancer Clin Oncol 20:947–954
3. Hamburger AW, Salmon SE (1977) Primary bioassay of human tumor stem cells. Science 197:461–463
4. Salmon SE, Hamburger AW, Soehnlen B, Durie BG, Alberts DS, Moon TE (1978) Quantitation of differential sensitivity of human-tumor stem cells to anticancer drugs. N Engl J Med 298:1321–1327
5. Twentyman PR (1985) Predictive chemosensitivity testing. Br J Cancer 51:295–299
6. Selby P, Buick RN, Tannock I (1983) A critical appraisal of the "human tumor stem-cell assay". N Engl J Med 308:129–134
7. Von Hoff DD (1983) Send this patient's tumor for culture and sensitivity. N Engl J Med 308:154–155

8. Pihl A, UICC Study Group on chemosensitivity testing of human tumors (1986) Problems—applications—future prospects. Int J Cancer 37:1–5

9. Singletary SE, Umbach GE, Spitzer G, Drewinko B, Tomasovic B, Ajani J, Hug V, Blumenschein G (1985) The human tumor stem cell assay revisited. Int J Cell Cloning 3:116–128

10. Baker FL, Spitzer G, Ajani JA, Brock WA, Lukeman J, Pathak S, Tomasovic B, Thielvoldt D, Williams M, Vines C et al (1986) Drug and radiation sensitivity measurements of successful primary monolayer culturing of human tumor cells using cell-adhesive matrix and supplemented medium. Cancer Res 46:1263–1274

11. Wilson AP, Ford CH, Newman CE, Howell A (1984) A comparison of three assays used for the in vitro chemosensitivity testing of human tumours. Br J Cancer 49:57–63

12. Smith HS, Lippman ME, Hiller AJ, Stampfer MR, Hackett AJ (1985) Response to doxorubicin of cultured normal and cancerous human mammary epithelial cells. J Natl Cancer Inst 74:341–347

13. Tofilon PJ, Buckley N, Deen DF (1984) Effect of cell-cell interactions on drug sensitivity and growth of drug-sensitive and -resistant tumor cells in spheroids. Science 226:862–864

14. Erlichman C, Vidgen D (1984) Cytotoxicity of adriamycin in MGH-U1 cells grown as monolayer cultures, spheroids, and xenografts in immune-deprived mice. Cancer Res 44:5369–5375

15. Zaffaroni N, Silvestrini R, Sanfilippo O, Daidone MG, Gasparini G (1985) In vitro activity of alkylating agents on human tumors as measured by a short-term antimetabolic assay. Tumori 71:555–561

16. Volm M, Wayss K, Kaufmann M, Mattern J (1979) Pretherapeutic detection of tumour resistance and the results of tumour chemotherapy. Eur J Cancer 15:983–993

17. Weisenthal LM, Marsden JA, Dill PL, Macaluso CK (1983) A novel dye exclusion method for testing in vitro chemosensitivity of human tumors. Cancer Res 43:749–757

18. Leighton J (1960) The propagation of aggregates of cancer cells: implications for therapy and a simple method of study. Cancer Chemother Rep 9:71–72

19. Leighton J, Kalla R, Turner JM Jr, Fennell RH Jr (1960) Pathogenesis of tumor invasion. II. Aggregate replication. Cancer Res 20:575–586

20. Leighton J (1959) Aggregate replication, a factor in the growth of cancer. Science 129(3347):466–467

21. Leighton J, Kalla R, Kline I, Belkin M (1959) Pathogenesis of tumor invasion. I. Interaction between normal tissues and transformed cells in tissue culture. Cancer Res 19(1):23–27

22. Dawe CJ, Potter M, Leighton J (1958) Progressions of a reticulum-cell sarcoma of the mouse in vivo and in vitro. J Natl Cancer Inst 21(4):753–781

23. Leighton J (1957) Contributions of tissue culture studies to an understanding of the biology of cancer: a review. Cancer Res 17(10):929–941

24. Kline I, Leighton J, Belkin M, Orr HC (1957) Some observations on the response of four established human cell strains to hydrocortisone in tissue culture. Cancer Res 17(8):780–784

25. Leighton J, Kline I, Belkin M, Legallais F, Orr HC (1957) The similarity in histologic appearance of some human cancer and normal cell strains in sponge-matrix tissue culture. Cancer Res 17(5):359–363

26. Leighton J, Kline I, Belkin M, Orr HC (1957) Effects of a podophyllotoxin derivative on tissue culture systems in which human cancer invades normal tissue. Cancer Res 17(4):336–344

27. Leighton J, Kline I, Belkin M, Tetenbaum Z (1956) Studies on human cancer using sponge-matrix tissue culture. III. The invasive properties of a carcinoma (strain HeLa) as influenced by temperature variations, by conditioned media, and in contact with rapidly growing chick embryonic tissue. J Natl Cancer Inst 16(6):1353–1373

28. Leighton J, Kline I, Orr HC (1956) Transformation of normal human fibroblasts into histologically malignant tissue in vitro. Science 123(3195):502

29. Leighton J (1954) The growth patterns of some transplantable animal tumors in sponge matrix tissue culture. J Natl Cancer Inst 15(2):275–293

30. Leighton J, Kline I (1954) Studies on human cancer using sponge matrix tissue culture. II. Invasion of connective tissue by carcinoma (strain HeLa). Tex Rep Biol Med 12(4):865–873

31. Leighton J (1954) Studies on human cancer using sponge matrix tissue culture. I. The growth patterns of a malignant melanoma, adenocarcinoma of the parotid gland, papillary adenocarcinoma of the thyroid gland, adenocarcinoma of the pancreas, and epidermoid carcinoma of the uterine cervix (Gey's HeLa strain). Tex Rep Biol Med 12(4):847–864

32. Leighton J (1951) A sponge matrix method for tissue culture; formation of organized aggregates of cells in vitro. J Natl Cancer Inst 12(3):545–561

33. Browning TH, Trier TS (1969) Organ culture of mucosal biopsies of human small intestine. J Clin Invest 48:1248

34. Schiff LJ (1975) Organ cultures of rat and hamster colon. In Vitro 11:46

35. Autrup H (1980) Explant culture of human colon. In: Methods in cell biology, vol 21B. Academic Press, New York, p 335

36. Autrup H (1983) In: Autrup H, Williams H (eds) Experimental colon carcinogenesis. CRC Press, Boca Raton

37. Sherwin RP, Richters A, Yellin AE, Donovan AJ (1980) Histoculture of human breast cancers. J Surg Oncol 13:9–20

38. Folkman J, Hochberg M (1973) Self-regulation of growth in three dimensions. J Exp Med 138(4):745–753

39. Carrel A (1912) On the permanent life of tissues outside of the organism. J Exp Med 15:516–528

40. Yoshida Y, Hillborn V, Hassett C, Melti P, Byers MJ, Freeman AE (1980) Characterization of mouse fetal lung cells cultured on a pigskin substrate. In Vitro 16:443–445

41. Yoshida Y, Hillborn V, Freeman AE (1980) Fine structure identification of organoid mouse lung cells cultured on a pigskin substrate. In Vitro 16:994–1006

42. Douglas WHJ, McAteer JA, Cavanagh T (1978) Organotypic culture of dissociated fetal rat lung cells on a collagen sponge matrix. Tissue Cult Assoc Manual 4:749–753

43. Vescio RA, Redfern CH, Nelson TJ, Ugoretz S, Stern PH, Hoffman RM (1987) *In vivo*-like drug responses of human tumors growing in three-dimensional gel-supported, primary culture. Proc Natl Acad Sci U S A 84:5029–5033

44. Chabner B (1982) Pharmacologic principles of cancer treatment. Saunders, Philadelphia

45. Alberts DS, Chen GHS (1980) Cloning of human tumor stem cells. Salmon SE. Liss, New York, pp 351–359

46. Hamilton E, Dobbin J (1982) [3H]thymidine labels less than half of the DNA-synthesizing cells in the mouse tumour, carcinoma NT. Cell Tissue Kinet 15:405–411

47. Scully RE (ed) (1979) Tumors of the ovary and maldeveloped gonads, Atlas of Tumor Pathology, 2nd Series Fascicle 16. Armed Forces Institute of Pathology, Washington, DC

48. Haskell CM (ed) (1985) Cancer treatment, 2nd edn. Saunders, Philadelphia

49. Freeman A, Hoffman RM (1986) *In vivo*-like growth of human tumors *in vitro*. Proc Natl Acad Sci U S A 83:2694–2698

50. Miller BE, Miller FR, Heppner GH (1981) Interactions between tumor subpopulations affecting their sensitivity to the antineoplastic agents cyclophosphamide and methotrexate. Cancer Res 41(11 Pt 1):4378–4381

51. Bhatia S, Frangioni JV, Hoffman RM, Iafrate AJ, Polyak K (2012) The challenges posed by cancer heterogeneity. Nat Biotechnol 30:604–610

52. Parikh RB, Schwartz JS, Navathe AS (2017) Beyond genes and molecules – A precision delivery initiative for precision medicine. N Engl J Med 376:1609–1612

Diagnosis and Pathological Analysis of Patient Cancers by Detection of Proliferating Cells in Gelfoam® Histoculture

Robert M. Hoffman

Abstract

Patient tumors grew in Gelfoam® histoculture with maintenance of tissue architecture, tumor-stromal interaction, and differentiated functions. In this chapter, we review the use of Gelfoam® histoculture to demonstrate proliferation indices of major solid cancer types explanted directly from surgery. Cell proliferation was visualized by histological autoradiography within the cultured tissues after [³H]thymidine incorporation by the proliferating cells. Epilumination polarization microscopy enables high-resolution imaging of the autoradiography of each cell. The histological status of the cultured tissues can be assessed simultaneously with the proliferation status. Carcinomas were observed to have areas of high epithelial proliferation with quiescent stromal cells. Sarcomas have high proliferation of the cancer cells of mesenchymal organ. Normal tissues can also proliferate at high rates. Mean growth fraction index (GFI) was highest for patient tumors with the pure subtype of small-cell lung cancer than other types of lung cancer.

Key words Gelfoam®, Three-dimensional histoculture, Tumors, Lung cancer, Small cell, Adenocarcinoma, Squamous-cell carcinoma, Autoradiography, Polarization, Microscopy, Proliferation index

1 Introduction

This chapter reviews the use of Gelfoam® three-dimensional histoculture of patient tumors to measure cell proliferation by autoradiography of any cell type within the tumor [1, 2] using high-resolution epiluminescence polarization microscopy. This assay can be used to calculate growth fraction indices or labeling indices that can distinguish tumor subtypes as well as indicate aggressiveness. We also demonstrate in many human tumor specimens, particularly those derived from colon cancer metastases, ovarian carcinoma, small-cell lung cancer and sarcoma, that the labeling indices in certain areas of the heterogeneous tumors can be extremely high, suggesting advanced malignancy [3].

Lung carcinomas are clinically divided into two major types: small-cell lung carcinoma (SCLC) [3] and non-small-cell lung

Robert M. Hoffman (ed.), *3D Sponge-Matrix Histoculture: Methods and Protocols*, Methods in Molecular Biology, vol. 1760, https://doi.org/10.1007/978-1-4939-7745-1_6, © Springer Science+Business Media, LLC, part of Springer Nature 2018

carcinoma (NSCLC) [4]. SCLC is more chemosensitive, in contrast to NSCLC, which includes adenocarcinomas, squamous cell carcinomas, and large-cell undifferentiated carcinomas of the lung, which are relatively chemoresistant and thus primarily treated with surgical resection for local disease [4]. Histological differences between these tumor types can be difficult to distinguish [5]. SCLC is further divided into three subtypes, "pure" small-cell, mixed small-cell/large cell, and combined small-cell carcinoma which may be difficult to distinguish [6, 7].

In this chapter, we demonstrate that the proliferation index of small-cell lung tumors in Gelfoam® histoculture differs significantly from those of non-small-cell lung cancer. Small-cell lung cancer has an elevated cell proliferation rate which can be used along with the histoculture growth patterns to reliably distinguish small-cell from non-small-cell lung cancer.

2 Materials

2.1 Culture Medium [2]

1. Culture medium (CM): RPMI 1640 with L-glutamine and phenol red, 100 µM modified Eagle's medium (MEM) nonessential amino acids (Thermo Fisher Scientific), 1 mM sodium pyruvate, 50 µg/ml gentamycin sulfate, 2.5 µg/ml amphotericin B (Thermo Fisher Scientific).

2. 10% Fetal bovine serum (FBS) (GIBCO): We recommend testing different batches of FBS before purchasing a large stock.

3. 200 mM L-glutamine (GIBCO).

2.2 Gelfoam® Sponge [2]

1. Gelfoam® Absorbable Collagen Sponge USP, 12–7 mm format (Pfizer, NDC: 0009-031508) as an example.

2.3 Materials for Preparation of Collagen Sponges [2]

1. Gelfoam® trimming aid: Double-edged blade.

2. Sterile Petri dish, 100 mm × 20 mm or wider.

3. Tissue transportation medium: Sterile phosphate buffer saline (PBS) or other solution with physiologic pH.

4. Sterile transportation container.

5. Sterile metal forceps or tweezers.

6. Sterile metal scissors.

7. Sterile flat weighing metal spatula.

8. 6-well culture plates.

9. Water-jacketed CO_2 incubator, set at 37 °C, 5% CO_2, ≥90% humidity.

10. Water bath.

2.4 Processing of Human Tumor Tissue [2]

1. Specimens of human tumors obtained from surgery.
2. Sterile Petri dish, 100 mm × 20 mm.
3. 70% ethanol solution.
4. Disinfectant solution for biological waste disposal.
5. Sterile forceps or tweezers.
6. Sterile scalpels and blades.
7. [³H]thymidine (2 µCi/mL; 1 Ci = 37 GBq) (Thermo Fisher Scientific, Chino, CA).
8. Kodak NTB-2 emulsion (Carestream Health, Rochester, NY).
9. Hematoxylin and eosin.

2.5 Evaluation of Results

1. Microscope with epi-polarization lighting system

---

3 Methods

1. Divide tumor into 1–2-mm diameter pieces and place on top of Gelfoam® previously hydrated in cell culture medium.
2. Add culture medium such that the upper part of the Gelfoam® is not covered.
3. Label cells within the three-dimensional histocultures capable of proliferation by administration of [³H]thymidine (2 µCi/mL; 1 Ci = 37 GBq) (17) for 4 days after 10–12 days in culture. Cellular DNA is labeled in any cells undergoing replication within the tissues.
4. After 4 days of labeling, wash the cultures with phosphate-buffered saline, place in histology capsules, and fix in 10% formalin.
5. Wash the cultures with phosphate-buffered saline, place in histology capsules, and fix in 10% formalin overnight.
6. Dehydrate the histocultures and rinse with paraffin.
7. Place the material in an embedding cassette and embed such that all six pieces of tumor are at the front of the paraffin block.
8. Section the block using a microtome so that a continuous ribbon of 5 µm sections is made, each containing six representative pieces of tumor.
9. Place approximately 10–15 sections of the ribbon on each side for autoradiographic processing and staining.
10. Make two slides of each condition.
11. Separate the slides, then coat with Kodak NTB-2 emulsion in a darkroom, and allow the slides to be exposed for 5 days at 4 °C before developing.

12. Stain the slides after rinsing with hematoxylin and eosin.

13. Analyze the slides by determining the percentage of cells undergoing DNA synthesis in treated vs. untreated tumor cultures, using a Nikon or an Olympus photomicroscope fitted with an epi-illumination polarization lighting system.

14. Determine replicating cells by the presence of bright green-reflecting silver grains over the cell nuclei that had incorporated [³H]thymidine. Consider cells labeled if ten or more grains are present over the cell nucleus. Exclude benign stromal cells from the measurement by their morphological appearance.

15. When counting, scan each slide at ×40 or ×100 power to locate the areas of maximum label. Once these areas are located, determine the percentage of cells undergoing DNA synthesis in at least three of the visual fields for each tumor piece at ×200.

16. Calculate the growth fraction index (GFI) by dividing the number of labeled cancer cells by the number of total cancer cells.

4 Results

Tumors and corresponding normal tissues were histocultured in Gelfoam® (see **Note 1**) for 14 days and then incubated with [³H] thymidine for days 11–14. Three-dimensional tissue organization representative of the original tissue was maintained throughout the culture period. Most tumors in Gelfoam® histoculture had at least some areas of high cellular proliferation and were intra-tumorally heterogeneous with regard to proliferation capability. Detection of radiolabeled proliferating cells was enhanced, as they exposed silver grains formed by autoradiography, by the scatter of incident polarized light [7, 8].

An example is the measure of the proliferation capacity of a metastatic colorectal tumor. High labeling in this culture was noted, where more than 90% of the cells in the most active region have proliferated during the labeling period of this relatively undifferentiated colon cancer metastasis to the liver (Fig. 1a). A high degree of cell proliferation was noted in the Gelfoam® histoculture of a small-cell lung cancer (Fig. 1b). In a Gelfoam® histoculture of an ovarian cancer, the epithelial cells are extensively proliferating and the stromal cells are quiescent (Fig. 1c). There was high proliferative capacity of an other ovarian carcinoma, in the cells which have invaded the supporting Gelfoam® (see **Note 2**). The invasive behavior of the tumor in Gelfoam® histoculture may mimic the way ovarian tumors frequently invade the peritoneal wall in vivo [8].

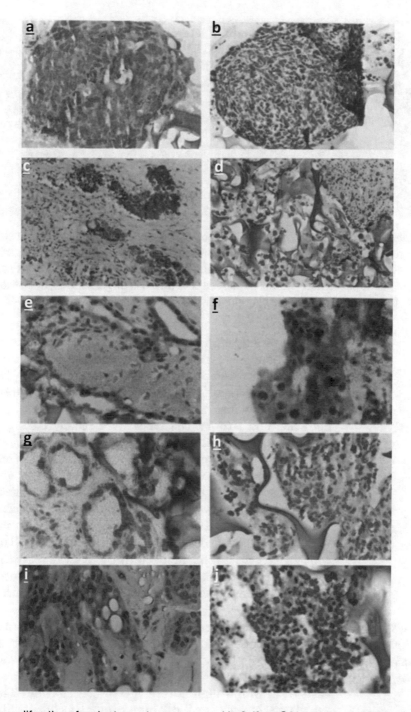

Fig. 1 Cellular proliferation of major tumor types measured in Gelfoam® histoculture in histological autoradio-grams analyzed with epi-illumination polarization microscopy. Tumors were in Gelfoam® histoculture for 14 days and were labeled with [³H]thymidine for the last 4 days. Cells with green grains over nuclei are radio-active and therefore proliferating. Original magnification was ×400 for all panels except for c and d, which were ×200. Final magnifications are ×3300 and ×1700, respectively. (**a**) Colon tumor; (**b**) small-cell lung carcinoma; (**c**) ovarian tumor; (**d**) ovarian tumor; (**e**) pancreas tumor; (**f**) bladder tumor; (**g**) kidney tumor; (**h**) brain tumor; (**i**) parotid tumor; and (**j**) Ewing's sarcoma [8]

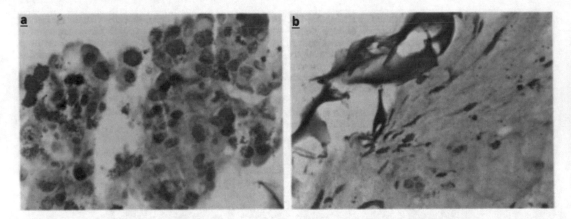

Fig. 2 Autoradiographic determination of proliferation of epithelial cells and normal stromal cells of a patient breast tumor grown in Gelfoam® histoculture. Original magnification was ×400; final magnification is ×3600. Proliferating epithelial cells are shown in **a** and proliferating stromal cells in **b**. Note that it is possible to distinguish, with regard to proliferation, between epithelial cells (which in this patient are, by histological criteria, malignant) and proliferating normal-appearing stromal cells [8]

The proliferation capacity of cancers in the pancreas, bladder, kidney, brain, and parotid gland, and a Ewing's sarcoma, was determined in Gelfoam® histoculture (Fig. 1d–j). Proliferating cells were observed in the glands of the tumors whose histology was exquisitely maintained in Gelfoam® histoculture. Proliferating epithelial and stromal cell types were readily distinguished. For example, in the breast tumor (Fig. 2a, b), the epithelial and stromal cells, were observed proliferating. Normal tissues in Gelfoam® histoculture can proliferate well. For example, tumor and adjacent normal tissue from the breast had extensive cell proliferation present (Fig. 3). There was a higher level of tissue organization maintained in the normal tissues. With polarized light without bright field, only the labeled cells are visualized (Fig. 4b). The image of the labeled cells can be digitized through a video camera (Fig. 4c) [8]. Greater than 90% of surgical specimens were cultured and analyzed for proliferative capacity in Gelfoam® histoculture [8].

4.1 Distinction of Small vs. Non-small-Cell Tumors by Growth Fraction Index (GFI)

Tumors were obtained from nine patients with pure small-cell lung cancer. All specimens came from the primary tumor and all nine tumors were successfully established in Gelfoam® histoculture. Large areas of tumor growth into the Gelfoam® were evident in each case, with a high percentage of cell labeling. The GFI was calculated for each small-cell lung tumor specimen with a mean of 79% (Fig. 5) [7].

Four additional tumors were identified as being of the mixed small-cell/large-cell subtype of SCLC by the criteria of the International Association for the Study of Lung Cancer [6]. These tumors were established in Gelfoam® histoculture and GFIs had a mean of 35% (SD ± 19%) [7].

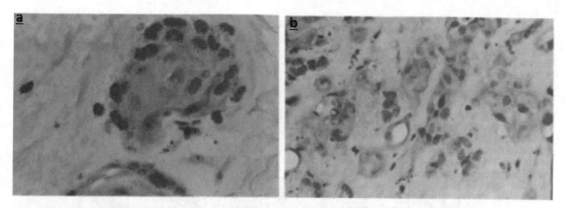

Fig. 3 Cellular proliferation in cancerous (**b**) and normal (**a**) breast tissues in Gelfoam® histoculture. Final magnification, ×3300. Note the relatively high level of proliferation in the stromal cells in the cancer tissue [8]

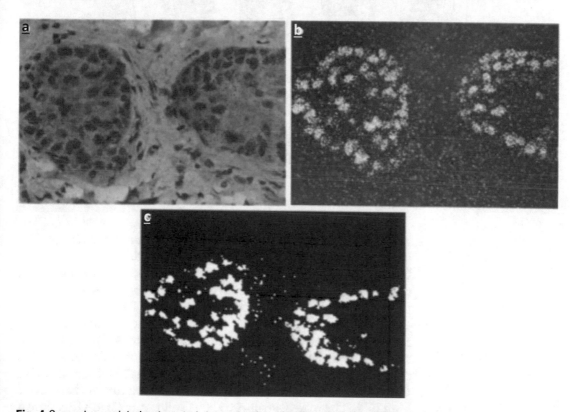

Fig. 4 Computer-assisted automated determination of cell proliferation indices of breast cancer in Gelfoam® histoculture using autoradiography, bright-field polarization microscopy, and image analysis. Final magnification, ×3300. (**a**) Bright-field and epi-illumination polarization microscopy of autoradiogram. Radioactive nuclei have exposed silver grains, which appear green due to polarization microscopy. (**b**) Epi-illumination polarization microscopy without bright-field light. Only dividing, autoradiographically-labeled, cells are visible. (**c**) Digitized processed image of **b** on computer monitor. Image represents only dividing, autoradiographically-labeled cells [8]. This is a very-high-resolution method to detect every proliferating cell of cancer or normal tissue in Gelfoam® histoculture

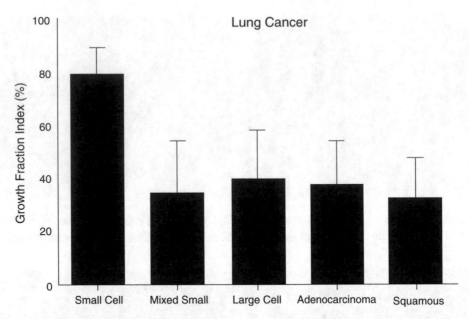

Fig. 5 GFIs were determined for 84 patient lung cancer in Gelfoam® by histological autoradiography. Tumors were cultured on Gelfoam® for 10 days and then labeled with [³H]dThd for 4 days. Autoradiography was performed on sectioned histocultures. The percentage of labeled cells in the regions of greatest labeling was determined and termed the growth fraction index (GFI). Mean GFI was 79 ± 10% for 9 patient tumors with the pure subtype of small-cell lung cancer. Mean GFIs were significantly lower for: 4 patient tumors with small-cell lung cancer of the mixed small-cell/large-cell subtype (35 ± 19%), 38 adenocarcinomas (38 ± 16%), 12 large-cell undifferentiated carcinomas (40 ± 18%), and 21 squamous cell carcinomas (33 ± 15%) ($P < 0.001$ in all cases) [7]

Tumors from 71 patients with non-small-cell lung cancer were explanted in Gelfoam histoculture. Culture success rate was 91%. GFIs had a mean of 37%. For squamous cell tumors the mean GFI was 33%. The mean GFI for large-cell tumors was 40%, whereas the mean GFI for adenocarcinomas was 38% [7].

The mean GFIs between the three classes of non-small-cell lung tumors were similar, ranging from 33 to 40%. This was in contrast to the mean GFI for small-cell tumors of 79% and was statistically significant ($P < 0.001$). All nine small-cell tumors had a GFI > 66%, whereas only 4 of 71 non-small-cell tumors had such a high GFI. The mixed small-cell/large cell tumors had GFIs similar to the NSCLCs with a significantly lower mean than that of the pure SCLCs ($P < 0.001$) [7].

4.2 Distinction of Small vs. Non-small-Cell Tumors by Growth Patterns

Figure 6a–c shows three different small-cell tumors grown in Gelfoam® histoculture and is fairly representative of the growth pattern obtained for this tumor type. The histocultured small-cell lung tumors frequently contained extensive regions of heavily labeled cells with infiltration into the Gelfoam® matrix. The pattern of SCLC tumor growth in Gelfoam® histoculture was

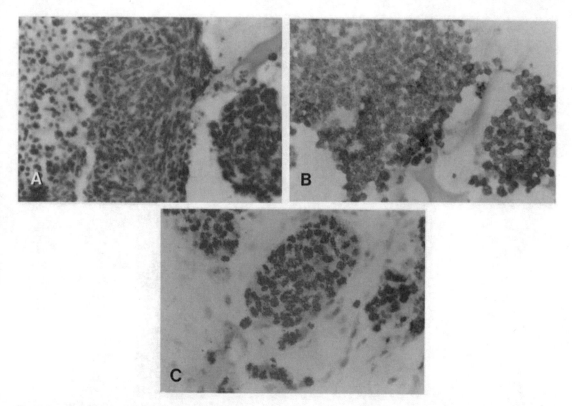

Fig. 6 Representative histological autoradiogram of tumor grown in Gelfoam® histoculture explants from three different patients with small-cell lung cancer (**a**, from patient 526; **b**, from patient 743; **c**, from patient 701). Tumor explants of 1 mm in size were in culture for 11 days followed by 4 days of labeling with [³H]thymidine (4 μCi/mL). After fixation, dehydration, and paraffin embedding, the histocultures were sectioned onto slides and coated with emulsion. After a 5-day exposure, the slides were stained with hematoxylin and eosin. Epi-illumination polarization microscopy at ×400 was used to identify replicating cells, depicted by bright green grains over cell nuclei. Small-cell tumors were characterized by consistent high cellularity and cell labeling as demonstrated in (**a**), (**b**), and (**c**) [7]

consistent in contrast to the wider phenotypic range of tumor growth observed for the non-small-cell lung tumors [7].

Most NSCLC in Gelfoam® histoculture had large areas of central necrosis with peripheral labeling in smaller clusters of cells. Although distinction between the three non-small cell types was not possible by the pattern of labeling and growth, certain characteristics of each tumor type seemed to be present. Large-cell tumors tended to grow into the Gelfoam® matrix peripherally and had the greatest cellularity of the NSCLC types, with many cells having a high cytoplasm-to-nuclear ratio (Fig. 7a–c). Adenocarcinomas of the lung frequently grew in duct-like patterns on the explant periphery with moderate cellularity within the tumor and moderate invasion into the Gelfoam® (Fig. 8a–c). Squamous cell tumors grew in swirls within the tumor explant, which often contained large areas of necrosis (Fig. 9a–c). Most of

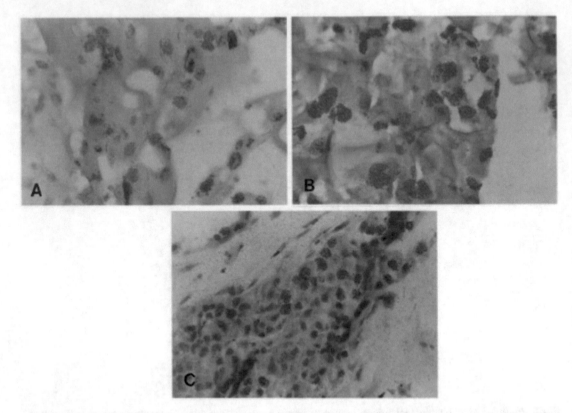

Fig. 7 Representative histological autoradiograms of tumor grown in Gelfoam® histoculture from three different patients with large-cell undifferentiated carcinoma of the lung (**a**, from patient 647; **b**, from patient 294; **c**, from patient 158). Large-cell undifferentiated carcinomas had, in general, the most cellularity of the non-small-cell lung tumors and tended to grow by invasion into the Gelfoam® matrix. Cell labeling in these areas of growth was much less than in comparative regions in small-cell lung tumors [7]

the cell labeling of squamous-cell tumors occurred in scattered cells far out in the Gelfoam® matrix. Some of the five bronchioalveolar cell tumors grew as a monocellular sheets after invading out of the Gelfoam® matrix onto the dish that matched the pattern of growth in vivo. The four mixed small-cell tumors had growth patterns more comparable to the NSCLCs. All nine pure small-cell tumors were correctly identified in photomicrographs alone by two separate blinded observers [7] (*see* **Notes 3** and **4**).

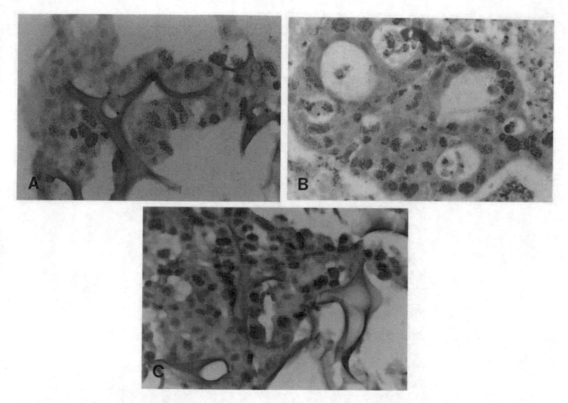

Fig. 8 Representative histological autoradiograms of tumors grown in Gelfoam® histoculture from three different patients with adenocarcinoma of the lung (**a**, from patient 159; **b**, from patient 193; **c**, from patient 496). Adenocarcinoma tumor grown in Gelfoam® histoculture showed the greatest cellularity and cell labeling in outgrowths from the tumor edge or bordering ducts within the tumor [7]

---

5 Notes

1. The ability to reproducibly culture human tumor explants in vitro has many uses for diagnosis, progression and treatment.

2. Prognostic information may be obtainable by measuring in vitro tumor proliferation characteristics can indicate aggressiveness.

3. Patterns of growth and invasion in Gelfoam® histoculture may be able to distinguish different tumor classes [7].

4. Invasive cells may be predominantly non-dividing [9].

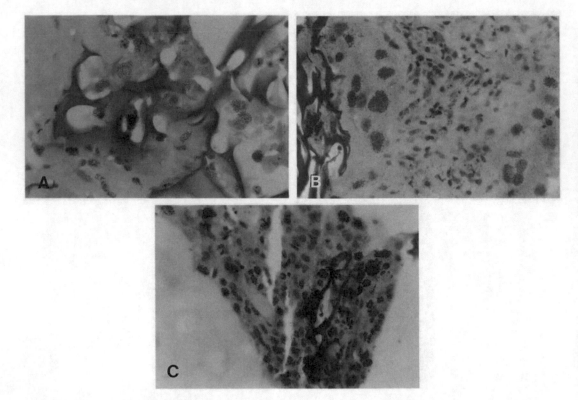

Fig. 9 Representative histological autoradiograms of tumors grown in Gelfoam® histoculture from three different patients with squamous-cell lung carcinomas (**a**, from patient 1043: **b**, from patient 369; **c**, from patient 502). Gelfoam® histocultured tumors from squamous-cell carcinomas frequently had large areas of central necrosis with low-cellularity monolayers along the tumor explant edge. Scattered cells deep in the Gelfoam® matrix were often present and usually labeled [7]

References

1. Freeman AE, Hoffman RA (1986) In vivo-like growth of human tumors in vitro. Proc Natl Acad Sci U S A 83:2694–2698

2. Vescio RA, Redfern CA, Nelson TJ, Ugoretz S, Stern PH, Hoffman RM (1987) In vivo-like drug responses of human tumors growing in three-dimensional gel-supported primary culture. Proc Natl Acad Sci U S A 84:5029–5033

3. Minna JD, Higgins GA, Glatstein EJ (1981) Cancer of the lung. In: DeVita VT, Hellman S, Rosenberg SA (eds) Principles and Practice of Oncology. J. B. Lippincott Co., Philadelphia, pp 396–474

4. Baylin SB, Weisburger WR, Eggleston JC, Mendelsohn G, Beaven M, Abeloff M, Ettinger D (1978) Variable content of histaminase. i.-dopa decarboxylase and calcitonin in small cell carcinoma of the lung. N Engl J Med 299:105–110

5. Feinstem AR, Gelfman NA, Yesner R, Auerbach O, Hackel DB, Pratt PC (1970) Observer variability in the histopathologic diagnosis of lung cancer. Am Rev Respir Dis 101:671–684

6. Hirsch FR, Matthews MJ, Aisner S, Campobasso O, Elema JD, Gazdar AF, Mackay B, Nasiell M, Shimosato Y, Steele RH, Yesner R, Zettergren L (1988) Histopathologic classification of small cell lung cancer. Cancer (Phila) 2:973–977

7. Vescio RA, Connors KM, Bordin GM, Robb JA, Youngkin T, Umbreit JN, Hoffman RM (1990) The distinction of small cell and non-small cell cancer by growth in native-state histoculture. Cancer Res 50:6095–6099

8. Hoffman RM, Monosov AZ, Connors KM, Herrera H, Price JH (1989) A general native-state method for determination of proliferation capacity of human normal and tumor tissues *in vitro*. Proc Natl Acad Sci U S A 86:2013–2017

9. Yano S, Miwa S, Mii S, Hiroshima Y, Uehara F, Yamamoto M, Kishimoto H, Tazawa H, Bouvet M, Fujiwara T, Hoffman RM (2014) Invading cancer cells are predominantly in G0/G1 resulting in chemoresistance demonstrated by real-time FUCCI imaging. Cell Cycle 13:953–960

Chapter 7

Clinical Correlation of the Histoculture Drug Response Assay in Gastrointestinal Cancer

Robert M. Hoffman

Abstract

The histoculture drug response assay (HDRA) with tumors histocultured on Gelfoam® was tested for clinical correlation for advanced gastric and colorectal cancer patients. In one study, 29 patients were treated with drugs shown to be ineffective in the HDRA, and all 29 cases showed clinical chemoresistance. In nine patients treated with drugs shown to be effective in the HDRA, six showed clinical chemoresponse and three showed arrest of disease progression. In a study of 32 patients with stage III and IV gastric cancer treated with mitomycin C and 5-fluorouracil (5-FU), the survival rate of 10 patients whose tumors were sensitive to either mitomycin C and/or 5-fluorouracil in the HDRA was significantly better than that of 22 patients whose tumors were insensitive to both drugs in the HDRA. Twenty-nine patients with stage III and IV colorectal cancer without remaining measurable tumor lesions after surgery were treated with fluoropyrimidines adjuvantly. The recurrence-free survival rate of 7 patients whose tumors were sensitive to 5-fluorouracil in the HDRA was significantly better than that of 22 patients whose tumors were insensitive in the HDRA. In a companion study of 128 gastric cancer patients whose tumors were evaluated in the HDRA, the overall and disease-free survival rates of the HDRA-sensitive group were found to be significantly higher than those of the HDRA-resistant group, treated with the same drugs.

Key words HDRA, Histoculture Drug Response Assay, Clinical correlation, Gastrointestinal cancer

1 Introduction

Heppner et al. [1–3] have demonstrated that the configuration of cells with respect to each other may affect their drug sensitivity, which suggested the idea that maintaining cancer cells in their native three-dimensional histological architecture may confer more accurate correlation to in vivo drug sensitivity. Folkman [4] has indicated that cells need to acquire their natural shape in order to have proper gene expression. Please see the Foreword and Chap. 2 of the present volume.

Hoffman and colleagues [1, 5–12] took advantage of the collagen sponge-gel matrix culture system developed by Leighton in the 1950s [13] to culture patient tumor tissue on Gelfoam® with maintenance of native tissue architecture. This approach was

Robert M. Hoffman (ed.), *3D Sponge-Matrix Histoculture: Methods and Protocols*, Methods in Molecular Biology, vol. 1760, https://doi.org/10.1007/978-1-4939-7745-1_7, © Springer Science+Business Media, LLC, part of Springer Nature 2018

termed histoculture by Sherwin et al. [14]. The critical importance of maintaining tumor architecture for accurate drug sensitivity determinations is reviewed by Hoffman [15–18]. Please see the Foreword and Chap. 2 of the present volume.

Hoffman and colleagues applied sponge-gel matrix histoculture of patient tumors to develop the histoculture drug response assay (HDRA) [12]. Please see Chap. 5 of the present volume.

Comparisons of drug-response spectra of human tumors in the HDRA and in nude mice showed that both drug resistance and sensitivity in the HDRA highly correlated to the in vivo response at approximately 90% [19]. A subsequent study with gastrointestinal cancer demonstrated that the HDRA correlated highly to historical clinical drug response [20].

The present chapter reviews correlative clinical trials comparing drug response in the HDRA and patients with gastrointestinal cancer.

2   Materials

1. Surgical knife.

2. Fine forceps.

3. Fine scissors.

4. Tissue and cell culture plate or flask: Tissue culture dishes (60 mm).

5. Culture medium (CM): RPMI1640 with L-glutamine and phenol red, 100 μM modified Eagle's medium (MEM) nonessential amino acids (Thermo Fisher Scientific), 1 mM sodium pyruvate, 50 μg/ml gentamycin sulfate, 2.5 μg/ml amphotericin B (Thermo Fisher Scientific), 15% fetal bovine serum (FBS). We recommend testing different batches of FBS before purchasing a large stock.

6. Gelfoam® Absorbable Collagen Sponge USP, 12–7 mm format (Pfizer, NDC: 0009-031508) as an example.

7. Gelfoam® trimming aid: Double-edged blade.

8. Sterile flat weighing metal spatula.

9. Water-jacketed CO_2 incubator, set at 37 °C, 5% CO_2, ≥90% humidity.

10. Water bath.

11. Tissue transportation medium: Sterile phosphate buffer saline (PBS) or other solution with physiologic pH.

12. Sterile transportation container.

13. 70% ethanol solution.

14. Disinfectant solution for biological waste disposal.

15. Collagenase type I (Sigma).

16. MTT (Sigma).

17. Sodium succinate (Wako Ind., Tokyo, Japan).

18. Dimethyl sulfoxide (DMSO) (Sigma).

19. Microplate reader (VersaMax, Sunnyvale, CA, USA).

20. [^{3}H]thymidine (4 µCi/ml; 1 Ci = 37 GBq).

21. Kodak NTB-2 emulsion (Carestream).

22. Phosphate-buffered saline (PBS).23. Microscope with an epi-polarization lighting system.

23. Microscope with an epi-polarization lighting system.

2.1 Drugs

1. Mitomycin C (Kyowa Hakko Kogyo Co., Ltd., Tokyo, Japan).

2. Doxorubicin (DOX) (Kyowa Hakko Kogyo Co., Ltd., Tokyo, Japan).

3. 5-Fluorouracil (5-FU) (Kyowa Hakko Kogyo Co., Ltd., Tokyo, Japan).

4. Cisplatinum (CDDP) (Bristol-Myers Squibb K.K., Tokyo, Japan).

5. UFT, a mixed compound of tegafur and uracil at a molar ratio of 1:4 (Taiho Pharmaceutical Co. Ltd., Tokyo, Japan).

3 Methods

3.1 Patients

Patients with advanced gastric cancer and advanced colorectal cancer were included in the study [6, 7].

3.2 HDRA with the MTT End Point

1. Scissor mince the cancerous portions of the specimens—minced into fragments approximately 0.5 mm in diameter.

2. Place tumor fragments on pre-hydrated Gelfoam® in 24-well plates.

3. Incubate the plates for 7 days at 37 °C.

4. Dissolve drugs in RPMI 1640 medium containing 20% FCS in a humidified atmosphere containing 95% air-5% CO$_2$.

5. Use the following cutoff concentrations of the drugs to distinguish in vitro sensitivity and resistance: 7.5 µg/ml for MMC; 15 µg/ml for DOX; 300 µg/ml for 5-FU; and 20 µg/ml for CDDP [21, 22].

6. After histoculture, add 100 µL 0.1 mg/ml collagenase and 100 µl MTT solution and incubate for another 8 h.

7. After extraction with DMSO, read the absorbance of the solution in each well at 540 nm.

8. Calculate the absorbance/mg of histocultured tumor tissue from the mean absorbance of tissue from four culture wells.

9. Calculate the inhibition rate using the following formula: inhibition rate (%) (1—mean absorbance of treated tumor/g/ mean absorbance of control tumor/g) × 100.

10. When the inhibition rate is 50% or more, score the chemosensitivity of tumors to drugs as positive [19, 20].

11. Test each drug concentration in at least three culture wells.

3.3 HDRA with the [³H]thymidine End Point (please see Chap. 5 in the present volume)

1. Alternatively, label cells in the histoculture with [³H]thymidine (4 µCi/mL; 1 Ci = 37 GBq) for three additional days after the drugs were removed. Cellular DNA is labeled in any cell undergoing replication within the tissues [7].

2. After 3 days of labeling, wash the cultures with PBS, place in histological capsules, and fix in 10% (v/v) formalin.

3. Dehydrate the cultures, embed in paraffin, section, and prepare for autoradiography using Kodak NTB-2 emulsion and counterstaining with hematoxylin and eosin [7].

4. Identify replicating cells by the presence of silver grains over their nuclei due to exposure of the NTB-2 emulsion to radioactive DNA.

5. Visualize the silver grains as bright green with an epipolarization lighting system [7].

6. Count the number of [³H]thymidine-labeled cells per field using ×200 magnification.

7. For each drug concentration, count one to three fields containing the highest number of labeled cells to identify the areas in the heterogenous tumor cultures having the least drug response.

8. Evaluate the control cultures in the same manner.

9. Evaluate two replicate cultures for each drug concentration to determine the in vitro response. Calculate percentage IR as 1—(treated/control value of [³H]thymidine-labeled cells) [7].

3.4 Patient Treatment

1. Treat patients who had remaining measurable tumor lesions after surgery with drugs which were shown to be effective in the HDRA of their individual tumors.

2. In cases where all four drugs tested showed negative antitumor activity in the HDRA, treat the patients with some combination of the four drugs.

3. Treat patients with advanced gastric and colorectal cancers without remaining measurable tumor lesions with postoperative adjuvant chemotherapy.

4. Exclude patients with stages I and II from the clinical correlation study because adjuvant chemotherapy of these patients is only occasionally performed at the discretion of the clinician since their expected prognosis is favorable.

5. Treat stage III and IV gastric cancer patients without remaining measurable tumor lesions with 30 mg/m² MMC and 400 mg/body/day UFT, a combination of tegafur and uracil in a molar ratio of 1:4 [21].

6. Treat stage III and IV colorectal cancer patients without remaining measurable tumor lesions adjuvantly with 400 mg/body/day UFT or 1-hexylcarbamoyl-5-FU [22].

3.5 Clinical Response to Chemotherapy

Patients with remaining measurable tumor lesions are eligible for retrospective evaluation of the assay results by comparison to the clinical effects of chemotherapy. Define complete response (CR) as the total disappearance of the tumor for at least 4 weeks. Define partial response (PR) as at least a 50% decrease in product of the longest perpendicular diameters of measurable lesions. Define no change as less than partial response without evidence of disease progression. Define progressive disease as any increase (>25%) in measurable lesions or the appearance of new lesions [6].

1. For patients without remaining measurable tumor lesions, use the survival rates and the recurrence-free survival rates determined by Kaplan-Meier analysis for the retrospective evaluation of correlation of the HDRA to efficacy of postoperative adjuvant chemotherapy [6].

4 Results

4.1 Evaluability of the Assay

One hundred and two of 107 gastric cancers and 106 of 109 colorectal cancers were found to be evaluable in the HDRA (96.3%). All 107 gastric and 109 colorectal tumor specimens weighed more than 150 mg and were applicable for the HDRA, resulting in a total evaluability rate of 96.3% (208/216) [6] (*see* **Notes 1** and **2**).

In a companion study, 172/206 (83.5%) of gastric cancer patient tumors could be evaluated in the HDRA [7], even though the tumor specimens were sent from Japan to San Diego.

4.2 Correlation of the Results of the HDRA and the Clinical Efficacy of Chemotherapy

The HDRA results and clinical efficacy of chemotherapy for each of 22 patients with gastric cancer and 16 with colorectal cancers with remaining measurable tumor lesions were correlated. For patients with gastric cancer, six patients whose tumors were sensitive to at least one of the drugs tested in the HDRA were treated with the HDRA-sensitive drugs. Three of these patients had drug responses (one CR and two PRs), while the remaining three showed stable disease. Sixteen patients whose tumors were insensitive to all four drugs tested in the HDRA were treated with some combination of the four drugs, and all of them showed drug resistance [6].

For patients with colon cancer, 5-FU was included in the treatment of all three patients who were sensitive to at least one of the drugs tested in the HDRA (including 5-FU). The 5-FU-based treatment was effective in all of these cases (one CR and two PRs). Thirteen patients whose tumors were insensitive to all four drugs tested in the HDRA were treated with some combination of the four drugs, and all of them showed drug resistance [6].

In total, there were 6 true-positive, 3 false-positive, and 29 true-negative cases, giving a correlation rate of the HDRA to the clinical effects of chemotherapy of 92.1% (35/38), with 66.7% (6/9) true-positive and 100% (29/29) true-negative rates, and 100% (6/6) sensitivity and 90.6% (29/32) specificity [6] (*see* **Notes 3** and **4**).

4.3 Correlation of the Results of the HDRA and the Efficacy of Postoperative Adjuvant Chemotherapy

There were 10 patients (group A) with advanced gastric cancer whose tumors were sensitive to MMC and/or 5-FU in the HDRA, and 22 patients (group B) whose tumors were insensitive to both MMC and 5-FU in the HDRA. There were no significant differences in terms of clinical and pathological characteristics between these two groups. Postoperative cancer recurrence was identified in three patients in group A, and one of them died of cancer. In group B, cancer recurrence was identified in 18 patients, and 16 of them died of cancer. The survival and recurrence-free survival rates evaluated according to Kaplan-Meier were significantly ($P < 0.005$ by log-rank test) better in group A than group B (Fig. 1) [6].

There were seven patients (group C) with stage III and IV colorectal cancer whose tumors were sensitive to 5-FU in the HDRA. Group D consisted of 22 patients whose tumors were insensitive to 5-FU in the HDRA. There were no significant differences in terms of clinical and pathological characteristics between these two groups. In group C, all seven patients were still alive without cancer recurrence. In group D, cancer recurrence was identified in eight patients, and three of them died of cancer. The survival and recurrence-free survival rates evaluated according to Kaplan-Meier were better in group C than group D with statistical significance ($P < 0.05$ by log-rank test) in the latter (Fig. 2) [6] (*see* **Note 5**).

In another study, there were 25 patients with advanced gastric cancer whose tumors were sensitive to MMC in the HDRA (sensitive group), and 98 patients whose tumors were resistant to MMC (resistant group) in the HDRA. There were no significant differences in terms of clinical and pathological characteristics between the resistant and sensitive groups. The overall survival rate evaluated according to Kaplan and Meier [14] was significantly better in the HDRA-sensitive group than in the

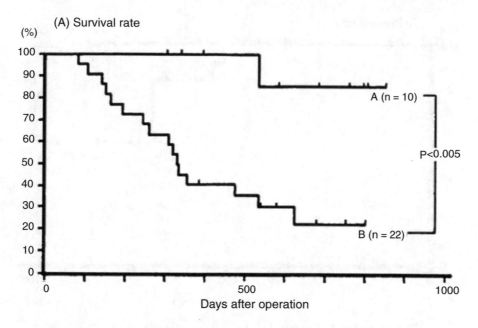

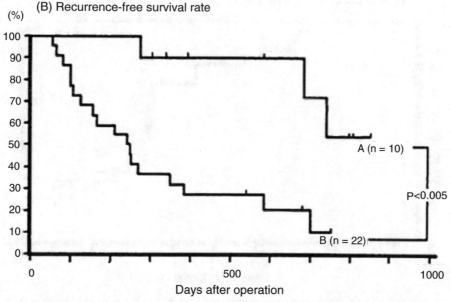

Fig. 1 (**a**) Kaplan-Meir survival curves; (**b**) recurrence-free survival curves following surgery of stage III and IV gastric cancer patients without remaining measurable tumor lesions who were treated adjuvantly with MMC and UFT. Group A consisted of ten patients whose tumors were sensitive to MMC and/or 5-FU in the HDRA. Group B consisted of 22 patients whose tumors were insensitive to both MMC and 5-FU in the HDRA. Survival rate and recurrence-free survival rate in group A were better than those in group B ($P < 0.005$ by log-rank test) [6]

HDRA-resistant group ($P = 0.02$ by the log-rank test with Bonferroni adjustment; Fig. 3) [7].

There were 20 advanced gastric cancer patients whose tumors were sensitive to 5-FU in the HDRA (sensitive group), and 99 patients whose tumors were resistant to 5-FU (resistant group) in the HDRA. There were no significant differences in terms of

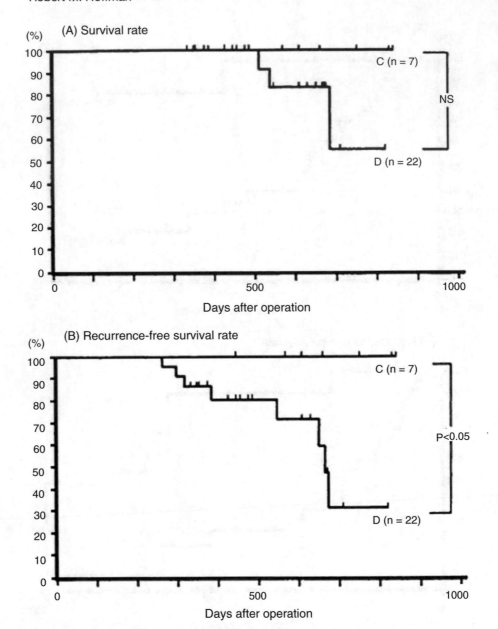

Fig. 2 (**a**) Kaplan-Meier survival curves; (**b**) recurrence-free survival curves following surgery of stage III and IV colorectal cancer patients without remaining measurable tumor lesions who were treated adjuvantly with fluoropyrimidines. Group C consisted of seven patients whose tumors were sensitive to 5-EU in the HDRA. Group D consisted of 22 patients whose tumors were insensitive to 5-FU in the HDRA. The recurrence-free survival rate in group C was better than that in group D (*P* < 0.05 by log rank test) [6]

clinical and pathological characteristics between the resistant and sensitive groups. The overall survival rate, evaluated according to Kaplan and Meier [23], was significantly better in the HDRA-sensitive group than in the HDRA-resistant group (*P* = 0.04 by log-rank test with adjustment: Fig. 4) [7].

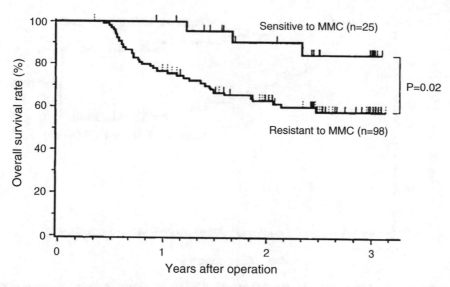

Fig. 3 Correlation of the overall survival rate of MMC- and UFT-treated stage III and IV gastric-cancer patients and the HDRA response of their tumors to MMC. The HDRA-sensitive group consisted of 25 patients whose tumors were sensitive to MMC in the HDRA. The resistant group consisted of 98 patients whose tumors were resistant to MMC in the HDRA. The overall survival rate of the HDRA-sensitive group was better than that of the HDRA-resistant group ($P = 0.02$ by log-rank test with adjustment) [7]

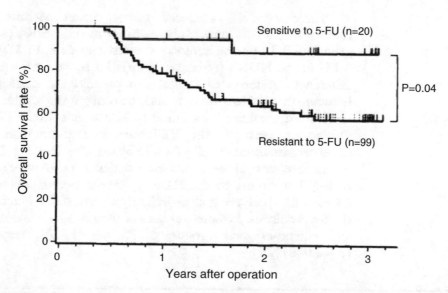

Fig. 4 Correlation of the overall survival rate of MMC- and UFT-treated stage III and IV gastric cancer patients and the HDRA response to 5-FU. The HDRA-sensitive group consisted of 20 patients whose tumors were sensitive to 5-FU in the HDRA. The HDRA-resistant group consisted of 99 patients whose tumors were insensitive to 5-FU. The overall survival rate of the sensitive group was better than that of the resistant group ($P = 0.04$ by log-rank test with adjustment) [7]

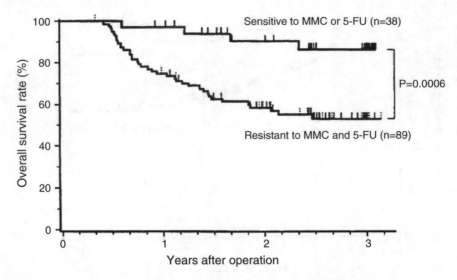

Fig. 5 Correlation of the overall survival rate of MMC- and UFT-treated stage III and IV gastric cancer patients and in the HDRA response of their tumors to MMC or 5-FU. The HDRA-sensitive group consisted of 38 patients whose tumors were sensitive to MMC or 5-FU in the HDRA. The HDRA-resistant group consisted of 89 patients whose tumors were insensitive to MMC and 5-FU in the HDRA. The overall survival rate of the sensitive group was better than that of the resistant group (P = 0.0006 by log-rank test) [7]

There were 38 advanced gastric cancer patients whose tumors were sensitive to MMC or 5-FU in the HDRA (sensitive group) and 89 patient tumors were resistant to both MMC and 5-FU in the HDRA (resistant group). There were no significant differences in terms of clinical and pathological characteristics between the HDRA-resistant and -sensitive groups. The overall survival rate evaluated according to Kaplan and Meier [23] was significantly better in the HDRA-sensitive group than in the HDRA-resistant group (P = 0.0006 by log-rank test; Fig. 5) [7].

In another study of colon cancer patients, 57.3% were sensitive to 5-FU according to the HDRA. After a median follow-up of 64 months, there was a statistically significant difference between the 5-year disease-free survival rates of the HDRA chemo-sensitive and chemo-resistant groups, 88.7% and 68.9%, respectively (p = 0.025) [24].

5   Notes

1. The HDRA demonstrated a very high rate of evaluability. This might be due to the fact that the HDRA allows cancer cells to maintain their native three-dimensional tissue architecture and viability [15, 16, 20].

2. The MTT end point allows chemosensitivity testing of resting cancer cells which cannot be evaluated by the [³H]thymidine

incorporation end point [19]. For chemosensitivity testing of gastrointestinal tumors, it seems that the MTT end point, which evaluates total tumor cell viability, is more appropriate than a proliferating-cell end point, since many cells in these types of tumors may be in the resting stage of the cell cycle [6, 25].

3. The very high true-negative rate demonstrated that the HDRA with the MTT end point is useful for the elimination of ineffective and harmful drugs for individual patients [6].

4. An important point of the HDRA appeared to be that there were no false-negative cases, despite the presence of some false-positive cases. All three false-positive cases were very advanced gastric cancers and had remaining unresectable local tumors, against which MMC was administered, resulting in no change and not in progressive disease, suggesting that the assay may have given useful information in these cases also. Therefore, the HDRA seems to be effective in identifying drug-sensitive patients as well as drug-resistant patients [6].

5. The most important point for in vitro chemosensitivity assays is to improve the patient's prognosis [26–28]. In this study, the HDRA was applied to post-operative adjuvant chemotherapy in cases without remaining measurable tumor lesions and demonstrated, although retrospectively, a potential use for selecting effective adjuvant cancer chemotherapy to improve the survival rates of the patients since in vitro drug sensitivity in the HDRA significantly correlated with patient survival [6].

References

1. Lawler EM, Miller FR, Heppner GH (1983) Significance of three-dimensional growth patterns of mammary tissues in collagen gel cultures. In Vitro 19:600–610

2. Miller BE, Miller FR, Heppner GH (1984) Assessing tumor drug sensitivity by a new in vitro assay which preserves tumor heterogeneity and subpopulation interactions. J Cell Physiol 3(Suppl):105–116

3. Miller BE, Miller FR, Heppner GH (1985) Factors affecting growth and drug sensitivity of mouse mammary tumor lines in collagen gel cultures. Cancer Res 45:4200–4205

4. Folkman J, Moscona A (1978) Role of cell shape in growth control. Nature 273(5661): 345–349

5. Vescio RA, Connors KM, Kubota T, Hoffman RM (1991) Correlation of histology and drug response of human tumors grown in native-state three-dimensional histoculture and in nude mice. Proc Natl Acad Sci U S A 88:5163–5166

6. Furukawa T, Kubota T, Hoffman RM (1995) Clinical applications of the histoculture drug response assay. Clin Cancer Res 1:305–311

7. Kubota T, Sasano N, Abe O, Nakao I, Kawamura E, Saito T, Endo M, Kimura K, Demura H, Sasano H, Nagura H, Ogawa N, Hoffman RM (1995) Potential of the histoculture drug response assay to contribute to cancer patient survival. Clin Cancer Res 1:1537–1543

8. Tanino H, Oura S, Hoffman RM, Kubota T, Furukawa T, Arimoto J, Yoshimasu T, Hirai I, Bessho T, Suzuma T, Sakurai T, Naito Y (2001) Acquisition of multidrug resistance in recurrent breast cancer demonstrated by the histoculture drug response assay. Anticancer Res 21:4083–4086

9. Singh B, Li R, Xu L, Poluri A, Patel S, Shaha AR, Pfister D, Sherman E, Hoffman RM, Shah J (2002) Prediction of survival in patients with head and neck cancer using the histoculture drug response assay. Head Neck 24:437–442

10. Jung PS, Kim DY, Kim MB, Lee SW, Kim JH, Kim YM, Kim YT, Hoffman RM, Nam JH (2013) Progression-free survival is accurately predicted in patients treated with chemotherapy for epithelial ovarian cancer by the histoculture drug response assay in a prospective correlative clinical trial at a single institution. Anticancer Res 33:1029–1034

11. Freeman A, Hoffman RM (1986) *In vivo*-like growth of human tumors *in vitro*. Proc Natl Acad Sci U S A 83:2694–2698

12. Vescio RA, Redfern CH, Nelson TJ, Ugoretz S, Stern PH, Hoffman RM (1987) *In vivo*-like drug response of human tumors growing in three-dimensional, gel-supported, primary culture. Proc Natl Acad Sci U S A 84:5029–5033

13. Leighton J (1951) A sponge matrix-method for tissue cultures. Formation of organized aggregates of cells in vitro. J Natl Cancer Inst 12:545–561

14. Sherwin RP, Richters A, Yellin AE, Donavan AJ (1980) Histoculture of human breast cancers. Surg Oncol 13:9–20

15. Hoffman RM (1991) In vitro sensitivity assays in cancer: A review, analysis and prognosis. J Clin Lab Anal 5:133–143

16. Hoffman RM (1993) In vitro assays for chemotherapy sensitivity. Crit Rev Oncol Hematol 15:99–111

17. Hoffman RM (1991) Three-dimensional histoculture: origins and application in cancer research. Cancer Cells 3:86–92

18. Hoffman RM (1993) To do tissue culture in two or three dimensions? That is the question. Stem Cells 11:105–111

19. Furukawa T, Kubota T, Watanabe M, Takahara T, Yamaguchi H, Takeuchi T, Kodaira S, Ishibiki K, Kitajima M, Hoffman RM (1992) High in vito-in vitro correlation of drug response using sponge gel-supported three-dimensional histoculture and the MTT end point. Int J Cancer 51:489–498

20. Furukawa T, Kubota T, Watanabe M, Kase S, Takahara T, Yamaguchi H, Takeuchi T, Teramoto T, Ishibiki K, Kitajima M, Hoffman RM (1992) Chemosensitivity testing of clinical gastrointestinal cancers using histoculture and the MiT end point. Anticancer Res 12:1377–1382

21. Maehara Y, Watanabe A, Kakeji Y, Baba H, Kohnoe S, Sugimachi K (1990) Postgastrectomy prescription of mitomycin C and UFT for patients with stage IV gastric carcinoma. Am J Surg 160:242–244

22. Koyama Y, Koyama Y (1980) Phase I study of a new antitumor drug, I-hexylcarbamoyl-5-fluorouracil (HCFU) administered orally: an HCFU clinical study group report. Cancer Treat Rep 64:861–867

23. Kaplan EL, Meier P (1958) Nonparametric estimation from incomplete observations. J Am Stat Assoc 53:457–481

24. Ji WB, Um JW, Ryu JS, Hong KD, Kim JS, Min BW, Joung SY, Lee JH, Kim YS (2017) Clinical significance of 5-fluorouracil chemosensitivity testing in patients with colorectal cancer. Anticancer Res 37:2679–2682

25. Yano S, Zhang Y, Miwa S, Tome Y, Hiroshima Y, Uehara F, Yamamoto M, Suetsugu A, Kishimoto H, Tazawa H, Zhao M, Bouvet M, Fujiwara T, Hoffman RM (2014) Spatial-temporal FUCCI imaging of each cell in a tumor demonstrates locational dependence of cell cycle dynamics and chemoresponsiveness. Cell Cycle 13:2110–2119

26. Von Hoff DD (1990) Selection of cancer chemotherapy for a patient by an in vitro assay versus a clinician. J Natl Cancer Inst 82:110–116

27. Gazdar AF, Steinberg SM, Russell EK, Linnoila RI, Oie HK, Ghosh BC, Cotelingham JD, Johnson BE, Minna JD, Ihde DC (1990) Correlation of in vitro drug-sensitivity testing results with response to chemotherapy and survival in extensive-stage small cell lung cancer: a prospective clinical trial. J Natl Cancer Inst 82:117–124

28. Leone LA, Meitner PA, Myers TJ, Grace WR, Gajewski WH, Fingert HJ, Rotman B (1991) Predictive value of the fluorescent cytoprint assay (FCA): a retrospective correlation study of in vitro chemosensitivity and individual responses to chemotherapy. Cancer Investig 9:491–503

Prospective Clinical Correlation of the Histoculture Drug Response Assay for Ovarian Cancer

Robert M. Hoffman, Phill-Seung Jung, Moon-Bo Kim, and Joo-Hyun Nam

Abstract

The histoculture drug response assay (HDRA) has been correlated clinically to a number of cancer types (please *see* Chaps. 7–11 of the present volume). The present chapter reviews the clinical trials of the HDRA for ovarian cancer. A prospective clinical trial of the HDRA for advanced epithelial ovarian cancer (AEOC) was performed at Asan Medical Center, Seoul, Korea. The clinical trial compared the efficacy of first-line therapy paclitaxel and carboplatinum in the HDRA and the clinical response for the patients whose tumors were tested in the HDRA. A series of patients (104) were treated with adjuvant combination chemotherapy of paclitaxel and carboplatinum after primary cytoreductive surgery. Tumor fragments were cultured on Gelfoam® and tested with paclitaxel and carboplatinum and evaluated with the MTT endpoint. Patients were categorized into two groups as either sensitive to both drugs (SS) or not sensitive to one or both drugs (R) based on HDRA results. The recurrence rate was much lower in the SS group compared to the R group, 29.2% vs. 69.8%, respectively. The SS group had a significantly longer progression-free survival compared to the R group, 34.0 months vs. 16.0 months, respectively. In another clinical trial, the HDRA was performed on 85 cases of ovarian cancer and 97% were evaluable. HDRA results were correlated to clinical response of 15 patients who received cisplatinum-based therapy that included doxorubicin and cyclophosphamide (CAP therapy). The true-positive rate was 88%, the true-negative rate was 86%, the sensitivity was 88%, the specificity was 86%, and the accurate prediction rate was 87% when HDRA results were compared to the response of the treated patients.

Key words Gelfoam® histoculture, HDRA, Ovarian cancer, Paclitaxel, Carboplatinum, Cisplatinum, Recurrence, Progression, Survival, Correlation, Prospective, Clinical trial

1 Introduction

First-line treatment of ovarian cancer involves cytoreductive surgery and taxane and platinum-based combination chemotherapy [1, 2] with a very high failure rate due to divergent responses to the chemotherapy [1]. Severe side effects often preclude the patient from further treatment after failure of the first-line regimen [2]. Thus, there is a need for individualized precision medicine for this disease. The histoculture drug response assay (HDRA) solves many previously encountered problems with in vitro testing of tumors as

Robert M. Hoffman (ed.), *3D Sponge-Matrix Histoculture: Methods and Protocols*, Methods in Molecular Biology, vol. 1760, https://doi.org/10.1007/978-1-4939-7745-1_8, © Springer Science+Business Media, LLC, part of Springer Nature 2018

it maintains three-dimensional tumor-tissue histology in culture and has a greater than 90% evaluation rate [3–11]. The HDRA has been reported to be clinically useful in breast, gastrointestinal, and head and neck cancer [12–17]. Please *see* Chaps. 7–11 in the present volume. Chemosensitivity determined by the HDRA was also a strong predictor of survival in patients with non-small-cell lung cancer [18–20].

This chapter reviews the clinical usefulness of the HDRA for ovarian cancer [21, 22].

2 Materials

1. Fine forceps.

2. Fine scissors.

3. Tissue plate.

4. Culture medium (CM): RPMI1640 with L-glutamine and phenol red, 100 μM modified Eagle's medium (MEM) nonessential amino acids (Thermo Fisher Scientific), 1 mM sodium pyruvate, 50 μg/ml gentamycin sulfate, 2.5 μg/ml amphotericin B (Thermo Fisher Scientific), 15% fetal bovine serum (FBS). We recommend testing different batches of FBS before purchasing a large stock.

5. Gelfoam® Absorbable Collagen Sponge USP, 12–7 mm format (Pfizer, NDC: 0009-031508) as an example.

6. Gelfoam® trimming aid: Double-edged blade.

7. Sterile flat weighing metal spatula.

8. Water-jacketed CO_2 incubator, set at 37 °C, 5% CO_2, ≥90% humidity.

9. Tissue transportation medium: Sterile phosphate buffer saline (PBS) or other solution with physiologic pH.

10. Sterile transport container.

11. Sterile scalpels and blades.

12. Disinfectant solution for biological waste disposal.

13. 70% ethanol solution.

14. Collagenase type I (Sigma).

15. Dimethyl sulfoxide (DMSO) (Sigma).

16. Hanks' balanced salt solution (HBSS; GIBCO, Gaithersburg, MD, USA).

3 Methods

3.1 Patients (Asan Hospital)

Patients (148) with epithelial ovarian cancer with International Federation of Gynecology and Obstetrics (FIGO) stage III–IV disease underwent primary cytoreductive surgery followed by combination chemotherapy of paclitaxel and carboplatinum between January 2007 and December 2012 at the Asan Medical Center (Seoul, Korea). All patients gave informed consent for HDRA testing. The study was approved by the Institutional Review Board of the Asan Medical Center (S2012-1977-0001). The median age of those patients was 58 years (36–80 years) and median follow-up duration was 26.0 months (1.5–66.0 months) [22].

3.2 Patients (Keio University)

The specimens were obtained from 88 patients with ovarian cancer treated surgically at the Keio University Hospital and related hospitals between April 1994 and July 1996 [21].

3.3 The Histoculture Drug Response Assay (Fig. 1)

1. Obtain tumor samples during cytoreductive surgery.

2. Aseptically wash tumors in Hanks' balanced salt solution (HBSS).

3. Mince into fragments approximately 0.5 mm in diameter.

4. Exclude necrotic and nonviable portions using 3-(4,5 dimethylthiazol-2-yl)-5-(3-carboxymethoxyphenyl)-2-(4-sulfo-phenyl)-2-tetrazolium (MTS; Sigma, St. Louis, MO, USA) staining.

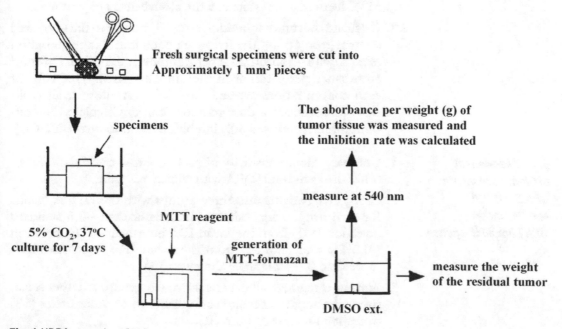

Fresh surgical specimens were cut into Approximately 1 mm³ pieces

specimens

The aborbance per weight (g) of tumor tissue was measured and the inhibition rate was calculated

measure at 540 nm

MTT reagent

5% CO₂, 37°C culture for 7 days

generation of MTT-formazan

measure the weight of the residual tumor

DMSO ext.

Fig. 1 HDRA procedure [21]

5. Weigh the viable cancer tissues on a chemical balance.

6. Place tumor fragments onto Gelfoam®.

7. Immerse Gelfoam® in 1 ml culture medium in 24-well plates.

8. Dissolve drugs at varying concentrations in Ham's F-12 medium (Gibco Rl), containing 20% heat-immobilized fetal calf serum.

9. Pipette the medium into each well (1 ml/well), taking care to avoid covering the Gelfoam® completely.

10. Use multiple concentrations of drugs in the HDRA, if possible.

11. Prepare four samples at each concentration, if possible.

12. Six and four replicates were concurrently run for the control and treatment groups, respectively.

13. Incubate for 72 h at 37 °C with 5% CO_2.

14. Treat the histocultures with 100 µl 0.06% collagenase type I (Sigma) in HBSS and 0.2% MTT (Sigma) in PBS containing 50 mM sodium succinate (Wako Ind., Tokyo, Japan).

15. Incubate plates for another 4 h, remove the medium, and add 0.5 mL dimethyl sulfoxide (DMSO) (Sigma) to each well to extract MTT formazan.

16. Transfer extracts from each well (100 µl) to a 96-well plate and measure the absorbance at 540 nm using a microplate reader (VersaMax, Sunnyvale, CA, USA).

17. Measure the weight of the residual tumor after extraction of MTT formazan and calculate the absorbance per g tumor.

18. Calculate the tumor inhibition rate (%), relative to the untreated control group using the following equation: tumor inhibition rate (%) = [1—(absorbance per g tumor in the treated group/ absorbance per g tumor in the untreated group)] × 100. At each concentration, average the inhibition rates for all replicates to construct a dose-response curve. Calculate the concentration that caused 50% inhibition of tumor growth (IC_{50}).

3.4 Assessment of Chemosensitivity with the HDRA and Corrlation with Clinical Response

1. Compare chemosensitivity of paclitaxel and carboplatinum or other drugs in the HDRA with clinical response.

2. Categorize patients into three groups with the HDRA results for each drug: drug resistant [IR greater than −0.5 standard deviation (SD) from the mean IR]; intermediate (IR between −0.5 SD and +0.5 SD from the mean IR); or sensitive (IR more than 0.5 + SD from the mean IR).

3. Further categorize all patients into two groups as either sensitive to both paclitaxel and carboplatinum or other drugs (SS) or resistant to one or both (R).

4. Analyze and compare inhibition rates for paclitaxel and carbo-platinum or other drugs and the decrease in the CA-125 level after the second and sixth cycles of chemotherapy, as well as progression-free survival (PFS), and overall survival (OS). Measure CA-125 levels before surgery and after every cycle of chemotherapy.

5. Perform imaging [(CT) or (PET/CT)] before surgery and after every three cycles of chemotherapy to examine the degree of disease control and define progression-free survival (PFS) as the interval between the date of debulking surgery and the date of imaging and in the patients without progression. Define PFS as the duration from the date of surgery to the last follow-up. Define overall survival (OS) as the interval between the date of debulking surgery and the date of death if the patient died of disease or the last follow-up date if the patient was still alive with or without disease.

6. Calculate accurate prediction rate (%) = (No. of true-positive cases + No. of true-negative cases)/Total no. of cases [23].

3.5 Statistical Analysis

1. Perform the χ^2 test and Fisher's exact test to compare the characteristics of patients and response rate using the Kaplan–Meier method for PFS and OS using SPSS ver.19.0 (SPSS Inc., Chicago, IL, USA) with a p-value less than 0.05 considered to be statistically significant [22].

4 Results and Discussion

In the Keio University clinical trial, the clinical response to CDDP-based chemotherapy was assessed in 15 cases whose residual tumor lesions were suitable for evaluation after surgery. The criteria of the Japanese Society of Cancer Therapy were employed in evaluating the direct effects of chemotherapy. The response rate, [percentage of cases showing complete response (CR) or partial response (PR)] was analyzed. Cases in which tumor tissue was determined by the HDRA to be sensitive and which the patient responded clinically to chemotherapy were regarded as true-positive (S/S) cases. Cases in which tumor tissue was determined by the HDRA to be resistant to chemotherapy and the patient did not respond to chemotherapy were regarded as true-negative (R/R) cases. Cases in which tumor tissue was determined by HDRA to be resistant but the patient responded to chemotherapy were regarded as false-negative (R/S) cases. Cases in which the tumor tissue was determined to be sensitive in the HDRA but the patient was clinically-resistant were regarded as false positive (S/R). The accurate prediction rate (%) was calculated using the following equation:

Accurate prediction rate (%) = (no. of true-positive cases + no. of true-negative cases)/total no. of cases [(23)].

In the Asan Hospital prospective study, during 26.0 (1.5–66.0) months of median follow-up, 56 patients (53.8%) experienced disease recurrence and 13 patients (12.5%) died of the disease. The mean inhibition rates for paclitaxel and carboplatinum were 46–58%, respectively [22].

For paclitaxel, 40 patients were classified as resistant, 24 classified as intermediate, and 40 classified as sensitive in the HDRA. For carboplatinum, 46 patients classified as resistant, 19 classified as intermediate, and 39 classified as sensitive in the HDRA. For carboplatinum, 24 patients were in the SS group and 49 in the RR group [22].

The median IR (range) for carboplatinum was significantly different in each group: 82% in the SS group vs. 45% in the RR group ($p < 0.01$). The median IR (range) for paclitaxel were also higher in the SS group than the R group: 66% (51–90%) vs. 35% (2–50%) but not statistically significant ($p = 0.643$) [22].

The recurrence rate was much lower in the SS group compared to the RR group: 29.2% vs. 71.4% ($p = 0.02$). The median PFS was also significantly longer in the SS group at 34.0 months compared to the RR group at 16.0 months ($p = 0.03$) [22] (*see* **Notes 1** and **2**).

The HDRA at Keio University was successfully carried out in a very high percentage of cases (97%) [21].

The HDRA yielded IC_{50} levels for cisplatinum (CDDP) which ranged from 6.4 to 132, with a mean 48.9 µg/ml. The sensitivity did not differ significantly with histologic type, clinical stage, or degree of tumor differentiation (data not shown) [21].

An approximate curve was obtained from this cumulative response curve using the curve fit method. Clinical response rates were applied to this equation to calculate the optimum cutoff IC_{50} level. The response rate of 33%, reported by Tobias and Griffiths et al. [24], was used as the historical clinical response rate to treatment with CDDP alone. The resultant optimum cutoff dose level for CDDP was determined to be 25 µg/ml. Therefore, cases for which the IC_{50} was below 25 µg/ml were scored as sensitive to CDDP [21].

Of the 88 cases on which HDRA was conducted, 67 received chemotherapy including CDDP after surgery. The sensitivity distribution did not depend on the degree of surgery. Of the 33 cases which received incomplete resection, 15 had residual lesions suitable for evaluation after surgery. For these 15 cases, the relationship between clinical responses and sensitivity in the HDRA was analyzed [21].

Of the eight cases in which the IC_{50} was determined by the HDRA to be 25 µg/ml or lower, two had complete response (CR), five had partial response (PR), and one had no change (NC). Of

Table 1

Correlation between clinical response and sensitivity in the HDRA [21]

Cutoff value (μg/mL)	Sensitivity with HDRA/clinical response			
	S/S	S/R	R/S	R/R
25	7	1	1	6

S sensitive, *R* resistant

Cases in which tumor tissue was determined by the HDRA to be sensitive and which the patient responded clinically to chemotherapy were regarded as true positive (S/S). Cases in which tumor tissue was determined in the HDRA to be resistant and which the patient did not respond to chemotherapy were regarded as true negative (R/R). Cases in which tumor tissue was determined by the HDRA to be sensitive but which the patient did not respond to chemotherapy were regarded as false positive (S/R). Cases in which tumor tissue was determined by the HDRA to be resistant but which the patient responded to chemotherapy were regarded as false-negative (R/S) cases [21]

the three cases with an IC_{50} between 25 had 70 µg/ml, one had CR and two had NC. Of the four cases with an IC_{50} over 70 µg/ml, all had NC [21].

Table 1 summarizes the results of the analysis of the relationship between clinical responses and sensitivity in the HDRA. When the cutoff level was set at 25 ug/ml (a level determined from the historical clinical response rate to CDDP), the true-positive rate was 88%, the true-negative rate was 86%, the sensitivity was 88%, the specificity was 86%, and the accurate prediction rate was 87% [21] (*see* **Notes 3, 4,** and **5**).

5   Notes

1. The clinical trial was performed prospectively in a single institution, Asan Medical Center, Seoul, Korea. Patients whose tumors showed higher IRs for both paclitaxel and carboplatinum in the HDRA had a significantly longer median PFS and lower recurrence rate. This finding indicates that the HDRA can be applied clinically when choosing initial chemotherapeutic agents after cytoreductive surgery for ovarian cancer and also for second-line agents in recurrent ovarian cancer. The results also showed that chemotherapy can be individualized by the HDRA for better prognosis by selecting more effective agents, especially when the IRs for first-line agents, carboplatinum and paclitaxel, are very low in the HDRA [22].

2. The median PFS appeared to be influenced significantly by carboplatinum but not by paclitaxel [22].

3. A number of drug sensitivity tests have been used to predict the responses of individuals to anticancer agents. However,

none of these tests satisfy the criteria of predicting clinical outcome, simplicity, and rapidity. The HDRA, which was used in this study, is based on three-dimensional culture. The most important characteristic of this method is that it maintains intercellular contact in a native three-dimensional architecture. The accurate response and a high rate of evaluability of the HDRA are most probably based on this feature [21].

4. The HDRA determined drug sensitivity in 7 days without causing any significant delay in the start of drug therapy. Furthermore, the HDRA evaluated a very high percentage (97%) of the cases [21].

5. The goal of a drug sensitivity assay is to accurately predict the response of tumors to chemotherapy not to attempt to be a scale model of the patient. Prospective studies will need to be carried out to determine to what degree the HDRA can improve clinical outcome in ovarian cancer.

References

1. Siegel R, Naishadham D, Jemal A (2012) Cancer statistics, 2012. CA Cancer J Clin 62:10–29

2. Joo WD, Lee JY, Kim JH, Yoo HJ, Roh HJ, Park JY, Kim DY, Kim YM, Kim YT, Nam JH (2009) Efficacy of taxane and platinum-based chemotherapy guided by extreme drug resistance assay in patients with epithelial ovarian cancer. J Gynecol Oncol 20:96–100

3. Vescio RA, Redfern CH, Nelson TJ, Ugoretz S, Stern PH, Hoffman RM (1987) In vivo-like drug responses of human tumors growing in three-dimensional gel-supported primary culture. Proc Natl Acad Sci U S A 84:5029–5033

4. Leighton J (1951) A sponge matrix method for tissue culture; formation of organized aggregates of cells in vitro. J Natl Cancer Inst 12:545–561

5. Hoffman RM (1991) In vitro sensitivity assays in cancer: a review, analysis, and prognosis. J Clin Lab Anal 5:133–143

6. Furukawa T, Kubota T, Watanabe M, Takahara T, Yamaguchi H, Takeuchi T, Kase S, Kodaira S, Ishibiki K, Kitajima M et al (1992) High in vitro-in vivo correlation of drug response using sponge-gel-supported three-dimensional histoculture and the MTT end point. Int J Cancer 51:489–498

7. Furukawa T, Kubota T, Watanabe M, Kase S, Takahara T, Yamaguchi H, Takeuchi T, Teramoto T, Ishibiki K, Kitajima M et al (1992) Chemosensitivity testing of clinical gastrointestinal cancers using histoculture and

the MTT end-point. Anticancer Res 12:1377–1382

8. Hoffman RM (1993) To do tissue culture in two or three dimensions? That is the question. Stem Cells 11:105–111

9. Hoffman RM (1993) In vitro assays for chemotherapy sensitivity. Crit Rev Oncol Hematol 15:99–111

10. Kubota T, Sasano N, Abe O, Nakao I, Kawamura E, Saito T, Endo M, Kimura K, Demura H, Sasano H et al (1995) Potential of the histoculture drug-response assay to contribute to cancer patient survival. Clin Cancer Res 1:1537–1543

11. Furukawa T, Kubota T, Hoffman RM (1995) Clinical applications of the histoculture drug response assay. Clin Cancer Res 1:305–311

12. Hoffman RM (2010) Histocultures and their use. In: Encyclopedia of Life Sciences. John Wiley and Sons, Ltd, Chichester. https://doi.org/10.1002/9780470015902.a0002573.pub2

13. Nakamura J, Imai E, Yoshihama M, Sasano H, Kubota T (1998) Histoculture drug response assay, a possible examination system for predicting the antitumor effect of aromatase inhibitors in patients with breast cancer. Anticancer Res 18:125–128

14. Kim JC, Kim DD, Lee YM, Kim TW, Cho DH, Kim MB, Ro SG, Kim SY, Kim YS, Lee JS (2009) Evaluation of novel histone deacetylase inhibitors as therapeutic agents for colorectal adenocarcinomas compared to established

regimens with the histoculture drug response assay. Int J Color Dis 24:209–218

15. Yoon YS, Kim CW, Roh SA, Cho DH, Kim GP, Hong YS, Kim TW, Kim MB, Kim JC (2012) Applicability of histoculture drug response assays in colorectal cancer chemotherapy. Anticancer Res 32:3581–3586

16. Singh B, Li R, Xu L, Poluri A, Patel S, Shaha AR, Pfister D, Sherman E, Hoffman RM, Shah J (2002) Prediction of survival in patients with head and neck cancer using the histoculture drug response assay. Head Neck 24:437–442

17. Hoffman RM (1991) Three-dimensional histoculture: origins and applications in cancer research. Cancer Cells 3:86–92

18. Yoshimasu T, Oura S, Maebeya S, Tanino H, Bessho T, Arimoto J, Sakurai T, Matsuyama K, Naito Y, Furukawa T, Yano T, Suzuma T, Hirai I (2000) Histoculture drug response assay on non-small-cell lung cancer. Gan To Kagaku Ryoho 27:717–722

19. Tamaki T, Oura S, Yoshimasu T, Ota F, Nakamura R, Shimizu Y, Kiyoi M, Naito K, Hirai Y, Okamura Y (2008) Histoculture drug response assay guided concurrent chemoradiotherapy for non-small cell lung cancer. Kyobu Geka 61:31–35

20. Tanahashi M, Yamada T, Moriyama S, Suzuki E, Niwa H (2008) The effect of the histoculture drug response assay (HDRA) based perioperative chemotherapy for non-small cell lung cancer. Kyobu Geka 61:26–30

21. Ohie S, Udagawa Y, Kozu A, Komuro Y, Aoki D, Nozawa S, Moossa AR, Hoffman RM (2000) Cisplatin sensitivity of ovarian cancer in the histoculture drug response assay correlates to clinical response to combination chemotherapy with cisplatin, doxorubicin and cyclophosphamide. Anticancer Res 20: 2049–2054

22. Jung PS, Kim DY, Kim MB, Lee SW, Kim JH, Kim YM, Kim YT, Hoffman RM, Nam JH (2013) Progression-free survival is accurately predicted in patients treated with chemotherapy for epithelial ovarian cancer by the histoculture drug response assay in a prospective correlative clinical trial at a single institution. Anticancer Res 33:1029–1034

23. Nakada S, Aoki D, Ohie S, Horiuchi M, Suzuki N, Kanasugi M, Susumu N, Udagawa Y, Nozawa S (2005) Chemosensitivity testing of ovarian cancer using the histoculture drug response assay: sensitivity to cisplatin and clinical response. Int J Gynecol Cancer 15:445–452

24. Tobias JS, Griffiths CT (1976) Management of ovarian carcinoma; current concepts and future prospects. N Engl J Med 294:818–822

Clinical Correlation of the Histoculture Drug Response Assay for Head and Neck Cancer

Robert M. Hoffman

Abstract

Gelfoam® histoculture was utilized to develop the histoculture drug response assay (HDRA) for head and neck cancer. Specimens of head and neck tumors were evaluated for sensitivity to the following drugs: cisplatinum (CDDP), 5-fluorouracil (5-FU), and the combination of CDDP and 5-FU. In the first clinical study at UCSD, 10 of 12 patients with tumors that were drug sensitive in Gelfoam® histoculture had either complete or partial response clinically. Comparisons of HDRA results, obtained with [³H]thymidine incorporation as the endpoint were made with clinical responses, i.e., complete response, partial response, or no response. The overall accuracy of the HDRA was 74% in this correlative clinical trial; the predictive positive value was 83%, the sensitivity was 71%, and the specificity was 78%. Seven of 11 patients with HDRA-resistant tumors demonstrated no response for a predictive negative value of 64%. In a subsequent study at Memorial Sloan Kettering Cancer Center, tumor specimens from 41 to 42 patients undergoing treatment for head and neck cancer were successfully evaluated by the HDRA. The histocultured tumors were treated with 5-FU and/or CDDP and a control group received no drug treatment. After completion of drug treatment, the relative cell survival in the tumors was determined using the MTT endpoint. Sensitivity was defined as a tumor inhibition rate (IR) of greater than 30%. Survival comparisons were performed using the generalized Wilcoxon test for the comparison of Kaplan-Meier survival curves. Resistance to 5-FU was observed in 13 cases (32%), to CDDP in 13 cases (32%), and to both agents in 11 cases (27%). The 2-year cause-specific survival was significantly greater for patients sensitive to 5-FU than patients who were resistant (85% vs. 64%), CDDP (86% vs. 64%), or both agents (85% vs. 63%). These results demonstrate the clinical usefulness of the HDRA for head and neck cancer.

Key words Gelfoam® histoculture, HDRA, Head and neck cancer, Clinical correlation, Cisplatinum (CDDP), 5-Fluorouracil (5-FU), [³H]thymidine, MTT

1 Introduction

The development of a reliable and practical assay to test sensitivity of individual tumors to various chemotherapy agents has been the focus of research by laboratories worldwide for many decades [1–13]. The in vitro tumor culture conditions should represent the complex microenvironment of the tumor [14].

Robert M. Hoffman (ed.), *3D Sponge-Matrix Histoculture: Methods and Protocols*, Methods in Molecular Biology, vol. 1760, https://doi.org/10.1007/978-1-4939-7745-1_9, © Springer Science+Business Media, LLC, part of Springer Nature 2018

Our approach has been the development of a three-dimensional (3D) Gelfoam® histoculture [15–28]. In this assay, important in vivo properties, including tissue architecture, are maintained over relatively long periods. The endpoint is the proportion of inhibition of cancer cell proliferation determined by histologic examination using autoradiographic techniques to determine [³H] thymidine incorporation or the biochemical MTT assay [14].

Squamous cell carcinomas of the upper aerodigestive tract have a mixed response to cisplatin-based chemotherapy regimens. Whereas some tumors completely regress clinically and pathologically, others show responses that are partial to nearly complete or minimal to none [29]. Clinical trials of patients with head and neck cancers have showed that only patients who have a complete response can survive [30]. Therefore it is imperative to predetermine active chemotherapy drugs for these patients. Trial and error with patients is not feasible [31].

We initially showed high take rate, success in maintaining the tumor histology, and the drug response of tumors derived from the head and neck in Gelfoam® histoculture [14].

Accordingly, HDRA assessment of chemoresponse was then tested as an outcome predictor in patients with head and neck cancer. There was a highly significant correlation of chemosensitivity in the HDRA and response and survival of patients with head and neck cancer [14, 32].

2   Materials

2.1   Gelfoam® Sponge

1. Sterile collagen sponges: Gelfoam® Absorbable Collagen Sponge USP, 12–7 mm format (Pfizer, NDC: 0009-0315-08).

2.2   Preparation of Gelfoam® Sponges

1. Gelfoam® trimming double-edge blade.

2. Sterile Petri dish, 100 mm × 20 mm or wider.

3. Culture medium (CM): RPMI1640 with L-glutamine and phenol red, 100 μM modified Eagle's medium (MEM) nonessential amino acids (Thermo Fisher Scientific), 1 mM sodium pyruvate, 50 μg/mL gentamycin sulfate, 2.5 μg/mL amphotericin B (Thermo Fisher Scientific), 15% fetal bovine serum (FBS). We recommend testing different batches of FBS before purchasing a large stock.

4. Sterile metal forceps or tweezers.

5. Sterile metal scissors.

6. Sterile flat weighing metal spatula.

7. 6-well culture plates.

8. Water-jacketed CO_2 incubator, set at 37 °C, 5% CO_2, ≥90% humidity.

9. Water bath.

2.3 Processing of Tumor Tissue

1. Hanks' balanced salt solution or sterile phosphate buffer saline (PBS) or cell culture medium for transporting tumor tissue.

2. Sterile transportation container.

3. Culture medium (please *see* above).

4. Timentin solution (100×): Add 100 mL of sterile water cell culture grade to a 3.1 *g* vial of Timentin® (GlaxoSmithKline, NDC: 0029-6571-26). Aliquot and store at −20 °C.

5. Sterile Petri dish, 100 mm × 20 mm.

6. Ethanol solution (70%).

7. Disinfectant solution for biological waste disposal.

8. Sterile forceps or tweezers.

9. Sterile scalpels and blades.

2.4 Drug Testing

1. 5-Fluorouracil (5-FU).

2. Cisplatinum (CDDP).

3. [³H]thymidine (2 µCi/mL; 1 Ci = 37 GBq) (Thermo Fisher Scientific, Chino, CA).

4. Kodak NTB-2 emulsion (Carestream Health, Rochester, NY).

5. Microscope with epi-polarization imaging system

6. MTT solution.

3 Methods

3.1 Patients

For the first HDRA feasibility study at UCSD, 11 patients had previously untreated disease, and 4 patients had recurrent disease. Staging for the 11 patients with untreated disease was as follows: stage I, one patient; stage II, no patients; stage III, five patients; and stage IV, five patients. Five specimens were obtained from punch biopsy and ten specimens were resected tumors. Thirteen specimens were taken from the primary disease site, and two were taken from metastatic lymph nodal disease. The primary sites of disease were as follows: larynx, three; oral tongue, two; unknown primary site, two; piriform sinus, two; base of tongue, one; tonsil, one; soft palate, one; oropharyngeal wall, one; parotid gland, one; and paranasal sinus, one. Thirteen specimens were histologically squamous cell carcinoma, one specimen was a poorly differentiated adenocarcinoma, and another was a poorly differentiated mucoepidermoid carcinoma [14].

The second study at UCSD comprised 26 patients with head and neck tumors who received cisplatinum (CDDP) chemotherapy for advanced untreated or recurrent disease. Fourteen patients had previously untreated disease, and 12 patients had recurrent disease. All patients with previously untreated disease had stage III–IV lesions; 21 patients had squamous cell carcinoma, two patients had adenocarcinoma, two patients had sarcoma, and one patient had a mucoepidermoid carcinoma. Sites of the primary disease included the following: oropharynx ($n = 9$), oral cavity ($n = 7$) paranasal sinuses ($n = 6$), larynx ($n = 1$), parotid gland ($n = 1$), unknown primary site ($n = 1$), and skin ($n = 1$). In four patients, specimens were obtained from cervical lymph node metastases, and in 22 patients, from the primary site [31].

Fifteen patients were treated with 120–200 mg/m^2 CDDP every 3–4 weeks. Eight patients received 150–200 mg/m^2 CDDP every 1–2 weeks with systemic sodium thiosulfate neutralization [31].

Tumor specimens from 41 head and neck cancer from Memorial Sloan-Kettering Cancer Center were successfully subjected to HDRA analysis. All patients had biopsy-proven squamous cell carcinoma. All patients treated with chemotherapy received CDDP in combination with 5-flourouracil (5-FU) [32].

3.2 HDRA Methods [31, 32]

1. Perform rigorous antibiotic washes of all resected tumor specimens, consisting of streptomycin, amikacin, penicillin, gentamicin, amphotericin β (Fungizone), chloramphenicol, and tetracycline

2. Place specimens immediately in transport media consisting of Hanks' balanced salt solution, L-glutamine (0.3 mg/ml), 10% fetal calf serum, nonessential amino acids (1:100 dilution of stock solution [obtained from Irvine Scientific, Irvine, CA]), and gentamicin (0.2 mg/ml).

3. Mince tumor into 1–2 mm diameter fragments and place on Gelfoam® as soon as possible hours after removal from the patient.

3.3 Drug Treatment

1. Add CDDP to the culture media at concentrations of 1.5 μg/ml (equivalent to the therapeutic plasma level), 15μg/ml (tenfold the therapeutic plasma level), and 37.5 μg/mL (25-fold the therapeutic plasma level).

2. Add 5-FU at concentrations of 4 μg/ml (therapeutic plasma level), 40 μg/ml (tenfold the therapeutic plasma level), and 100 μg/ml (25-fold the therapeutic plasma level).

3. Add combinations of CDDP and 5-FU in doses corresponding to the therapeutic level and tenfold the respective individual doses for each agent [14].

4. Dissolve the chemotherapeutic agents in DMEM/Ham's F12 medium with 10% FCS and gentamicin [32].

5. Expose histocultured tumors to the drugs dissolved in the culture media for 24 h on day 1.

3.4 Determination of Cell Proliferation

1. After drug treatment, add 4 μCi/mL of [³H]thymidine (1 Ci = 37 GBq) to label replicating cells for 4 additional days [14].

3.5 Biochemical Activity

1. After completion of drug treatment, determine the relative amount of cellular activity using the solution containing 3-(4,5-dimethylthiazol-2yl)-2,5-diphenyl-2H tetrazolium bromide (MTT) at a concentration of 0.4 mg/ml and incubate for 6 h and read the optical density or fluorescence [32].

3.6 Histological Preparation

1. Wash the histocultured tumors with isotonic phosphate-buffered saline, place in histology capsules, and fix in 10% (vol/vol) formaldehyde solution.

2. Dehydrate the histocultures.

3. Wash, embed in paraffin, section, and place onto slides.

4. Deparaffinize the slides and coat with an emulsion (Kodak NTB2) in a darkroom and expose for 5 days at 4 °C before developing.

5. After rinsing, stain the slides with hematoxylin-eosin [14].

6. Analyze the slides for [³H]thymidine incorporation with a polarizing microscope (at ×400 power).

7. Identify the replicating cells by the presence of bright-green-reflecting silver grains over the cell nuclei.

8. Determine the percentage of cells undergoing DNA synthesis in at least three visual fields with the heaviest labeling for each tumor piece.

9. Calculate the growth fraction index by dividing the number of labeled cancer cells by the number of unlabeled cancer cells (Fig. 1) [14].

10. Define chemosensitivity as a tumor inhibition rate (IR) of greater than 30% using either [³H]thymidine incorporation or MTT values [32].

11. Use a two-tailed p-value of less than or equal to 0.05 to accept significance to summarize study data.

12. Perform nonparametric qualitative and quantitative comparisons using the Fisher's exact test and Mann-Whitney U test, respectively.

13. Generate survival curves of patient tumors using the Kaplan-Meier method.

14. Perform survival comparisons using the generalized Wilcoxon test.

15. Perform multivariate analysis using the Cox regression model [32].

4 Results

In the initial feasibility study, five of the ten specimens were sensitive to CDDP in the HDRA, including one specimen with poorly differentiated adenocarcinoma and one specimen with poorly differentiated mucoepidermoid carcinoma, which was completely or partially sensitive to this drug. Four of the nine specimens were also sensitive to 5-FU. Three specimens were completely sensitive to the combination of CDDP and 5-FU, four were partially sensitive, and one was resistant [14] (*see* **Note 1**).

In the UCSD clinical trial, the in vitro results correlate with clinical response, partial responses, and complete responses in 17 (74%) of 23 cases. There were seven non-responders which accurately correlated to HDRA results and four cases with false-negative results for a predictive negative value of 7/11 or 64%. The sensitivity of the HDRA in this clinical trial correlation was 71%, and the specificity was 78%. It should be noted that in 12 of the patients, the biopsy specimens were taken after the patients had been treated with CDDP. The pretreated patients and nonpretreated patients were compared, and the overall accuracy of the HDRA was almost exactly the same, i.e., 75% vs. 73%, respectively [31] (*see* **Notes 2–4**).

In the Memorial Sloan Kettering Cancer Center study, 41 of 42 cases (98%) were successfully histocultured and analyzed for response to 5-FU and CDDP. One case was excluded because of the development of bacterial infection in the histoculture [31]. Resistance to 5-FU in the HDRA was present in 13 cases (32%), to CDDP in 13 cases (32%), and to both agents given simultaneously in 11 cases (27%) [32]. There were no statistical differences in patient, tumor, or treatment characteristics between resistant and sensitive cases. Recurrent cancers were not more chemoresistant than the primary tumor [32] (*see* **Notes 5** and **6**).

Significant differences in outcome were noted between resistant and sensitive patients. Based on the HDRA assessment, the 2-year cause-specific survival was significantly better for cases sensitive to 5-FU (85% vs. 64%; $p = 0.04$), sensitive to CDDP (86% vs. 64%; $p = 0.05$), or sensitive to both agents (85% vs. 63%; $p = 0.01$) [31] (Figs. 1 and 2).

These studies indicate that the HDRA can accurately determine optimal drug treatment for individual patients with head and neck cancer.

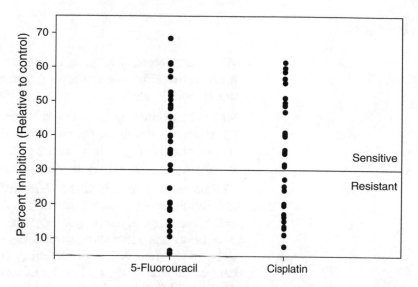

Fig. 1 Degree of inhibition determined by the histoculture drug response assay in 41 patients. The inhibition rate was calculated relative to a control group. A rate of 30% or less was deemed resistant [32]

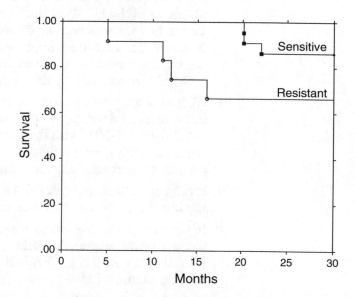

Fig. 2 Cause-specific survival by multidrug resistance based on the histoculture drug response assay. Multidrug resistance was defined as in vitro resistance to both CDDP and 5-FU. Survival comparisons were performed using the Wilcoxon test ($p = 0.01$) [32]

5 Notes

1. We were successful in maintaining the cultures to study the proliferative effects of a variety of agents after instituting rigorous antibiotic washes.

2. Since chemotherapy has become an important investigational treatment modality for patients with advanced head and neck cancer, the development of assays predictive of chemosensitivity could provide useful guidelines for this purpose [14].

3. The potential strength of the Gelfoam® histoculture assay lies in its ability to overcome many of the problems associated with other techniques. Its major advantage is its morphological endpoint, one that uses autoradiography and histology to determine whether the proliferating cells are actually cancer cells or stromal cells, such as lymphocytes or fibroblasts. The Gelfoam® HDRA is also capable of measuring drug effects on nonproliferating cancer cells. Although clonogenic cells are thought to be the most critical for tumor growth, there is evidence that resting cells can be recruited at some later time to become replicating or dividing cells [27]. Another theoretical advantage of the HDRA is the preservation of the tissue architecture in a three-dimensional configuration. The cancer cell-stromal cell interaction is important for control of cell growth and differentiation through autocrine and paracrine pathways, is, thereby, maintained by the culture configuration [14, 40].

4. We have attempted to overcome the major problems of clonogenic assays by developing a Gelfoam® HDRA for solid tumors [14, 17–19, 33–39]. With this system, important in vivo properties, including three-dimensional tissue architecture, are maintained over relatively long time periods [19, 31–39].

5. Ideally, an in vitro assay should be accurate in predicting both sensitivity and resistance to chemotherapy agents.

6. In vitro assays for chemoresponsiveness may serve as an important surrogate prognostic marker. The patients whose tumors were sensitive in the HDRA had significantly better survival than the controls [31, 32].

References

1. Hamburger AW, Salmon SE (1977) Primary bio-assay of human tumor stem cells. Science 197:461–463

2. Courtenay VD, Selby PJ, Smith IE, Mills J, Peckham MJ (1978) Growth of human tumour cell colonies from biopsies using two soft-agar techniques. Br J Cancer 38:77–81

3. Sondak VJ, Korn EL, Korn DT (1988) In vitro testing of chemotherapeutic combinations in a rapid thymidine incorporation assay. Int J Cell Cloning 6:378–391

4. Selby PJ, Steel GG (1982) Use of agar diffusion chamber for the exposure of human tumor cells to drugs. Cancer Res 42:4756–4762

5. Maurer HR, Ali-Osman F (1981) Tumor stem cell cloning in agar-containing capillaries. Naturwissenschaften 62:381–383

6. Von Hoff DD, Forseth BJ, Huong M, Buchok JB, Lathan B (1986) Improved plating efficiencies for human tumors cloned in capillary tubes versus Petri dishes. Cancer Res 46:4012–4017

7. Baker FL, Spitzer G, Ajani JA, Brock WA, Lukeman J, Pathak S, Tomasovic B, Thielvoldt D, Williams M, Vines C, Tofilon P (1986) Drug and radiation sensitivity measurements of primary monolayer culture of human tumor cells using cell-adhesive matrix and supplemental medium. Cancer Res 46:1263–1274

8. Daidone MG, Silvestrini R, Sanfilippo O, Zaffaroni N, Varini M, De Lena M (1985) Reliability of an in vitro short-term assay to predict the drug sensitivity of human breast cancer. Cancer 56:450–456

9. Weisenthal LM, Marsden JA, Dill PL, Macaluso CK (1983) A novel dye exclusion method for testing in vitro chemosensitivity of human tumors. Cancer Res 43:749–757

10. Halfter K, Mayer B (2017) Bringing 3D tumor models to the clinic–predictive value for personalized medicine. Biotechnol J 12:n/a. https://doi.org/10.1002/biot.201600295

11. Ji WB, Um JW, Ryu JS, Hong KD, Kim JS, Min BW, Joung SY, Lee JH, Kim YS (2017) Clinical significance of 5-fluorouracil chemosensitivity testing in patients with colorectal cancer. Anticancer Res 37:2679–2682

12. Karekla E, Liao WJ, Sharp B, Pugh J, Reid H, Quesne JL, Moore D, Pritchard C, MacFarlane M, Pringle JH (2017) Ex vivo explant cultures of non-small cell lung carcinoma enable evaluation of primary tumor responses to anticancer therapy. Cancer Res 77:2029–2039

13. Yoon YS, Kim CW, Roh SA, Cho DH, Kim TW, Kim MB, Kim JC (2017) Development and applicability of integrative tumor response assays for metastatic colorectal cancer. Anticancer Res 37:1297–1303

14. Robbins KT, Varki NM, Storniolo AM, Hoffman H, Hoffman RM (1991) Drug response of head and neck tumors in native-state histoculture. Arch Otolaryngol Head Neck Surg 117:83–86

15. Freeman A, Hoffman RM (1986) In vivo-like growth of human tumors in vitro. Proc Natl Acad Sci U S A 83:2694–2698

16. Vescio RA, Redfern CH, Nelson TJ, Ugoretz S, Stern PH, Hoffman RM (1987) In vivo-like drug response of human tumors growing in three-dimensional, gel-supported, primary culture. Proc Natl Acad Sci U S A 84:5029–5033

17. Hoffman RM, Monosov AZ, Connors KM, Herrera H, Price JH (1989) A general native-state method for determination of proliferation capacity of human normal and tumor tissues in vitro. Proc Natl Acad Sci U S A 86:2013–2017

18. Hoffman RM (1991) Three-dimensional gel-supported native-state histoculture for evaluation of tumor-specific pharmacological activity: principles, practices and possibilities. J Cell Pharmacol 2:189–201

19. Guadagni F, Roselli M, Hoffman RM (1991) Maintenance of expression of tumor antigens in three-dimensional in vitro human tumor gel-supported histoculture. Anticancer Res 11:543–546

20. Furukawa T, Kubota T, Watanabe M, Takahara T, Yamaguchi H, Takeuchi T, Kase S, Kodaira S, Ishibiki K, Kitajima M, Hoffman RM (1992) High in vitro-in vivo correlation of drug response using sponge-gel-supported three-dimensional histoculture and the MTT end point. Int J Cancer 51:489–498

21. Geller J, Partido C, Sionit L, Youngkin T, Nachtsheim D, Espanol M, Tan Y, Hoffman RM (1997) Comparison of androgen-independent growth and androgen-dependent growth in BPH and cancer tissue from the same radical prostatectomies in sponge-gel matrix histoculture. Prostate 31:250–254

22. Hoffman RM (1994) Three-dimensional sponge-gel matrix histoculture of human tumors: methods and applications. In: Celis J (ed) Cell biology: a laboratory handbook. Academic Press, San Diego, pp 367–379

23. Hoffman RM (1998) Three-dimensional sponge-gel matrix histoculture of human tumors: methods and applications. In: Celis J (ed) Cell biology: a laboratory handbook, vol vol. 1, 2nd edn. Academic Press, San Diego, pp 377–389

24. Mii S, Uehara F, Yano S, Tran B, Miwa S, Hiroshima Y, Amoh Y, Katsuoka K, Hoffman RM (2013) Nestin-expressing stem cells promote nerve growth in long-term 3-dimensional Gelfoam®-supported histoculture. PLoS One 8:e67153

25. Tome Y, Uehara F, Mii S, Yano S, Zhang L, Sugimoto N, Maehara H, Bouvet M, Tsuchiya H, Kanaya F, Hoffman RM (2014) 3-dimensional tissue is formed from cancer cells in vitro on Gelfoam®, but not on Matrigel™. J Cell Biochem 115:1362–1367

26. Zhang L, Wu C, Bouvet M, Yano S, Hoffman RM (2015) Traditional Chinese medicine herbal mixture LQ arrests FUCCI-expressing

HeLa cells in G0/G1 phase in 2D plastic, 2.5D Matrigel®, and 3D Gelfoam® culture visualized with FUCCI imaging. Oncotarget 6: 5292–5298

27. Yano S, Miwa S, Mii S, Hiroshima Y, Uehara F, Kishimoto H, Tazawa H, Zhao M, Bouvet M, Fujiwara T, Hoffman RM (2015) Cancer cells mimic *in vivo* spatial-temporal cell-cycle phase distribution and chemosensitivity in 3-dimensional Gelfoam® histoculture but not 2-dimensional culture as visualized with real-time FUCCI imaging. Cell Cycle 14:808–819

28. Yano S, Takehara K, Miwa S, Kishimoto H, Tazawa H, Urata Y, Kagawa S, Bouvet M, Fujiwara T, Hoffman RM (2017) GFP labeling kinetics of triple-negative human breast cancer by a killer-reporter adenovirus in 3D Gelfoam® histoculture. In Vitro Cell Dev Biol Anim 53:479–482

29. Jacobs JR, Pajak TF, Kinzie J, Al-Sarraf M, Davis L, Hanks GA, Weigensberg I, Leibel S (1987) Induction chemotherapy in advanced head and neck cancer. Arch Otolaryngol Head Neck Surg 113:193–197

30. Cognetti F, Pinnaro P, Ruggeri EM, Carlini P, Perrino A, Impiombato FA, Calabresi F, Chilelli MG, Giannarelli D (1989) Prognostic factors for chemotherapy response and survival using combination chemotherapy as initial treatment of advanced head and neck squamous cell cancer. J Clin Oncol 7:829–837

31. Robbins KT, Connors KM, Storniolo AM, Hanchett C, Hoffman RM (1994) Sponge-gel-supported histoculture drug-response assay for head and neck cancer. Correlations with clinical response to cisplatin. Arch Otolaryngol Head Neck Surg 120:288–292

32. Singh B, Li R, Xu L, Poluri A, Patel S, Shaha AR, Pfister D, Sherman E, Hoffman RM, Shah J (2002) Prediction of survival in patients with head and neck cancer using the histoculture drug response assay. Head Neck 24:437–442

33. Wallen JW, Cate RL, Kiefer DM, Riemen MW, Martinez D, Hoffman RM, Donahoe PK, Von Hoff DD, Pepinsky B, Oliff A (1989) Minimal antiproliferative effect of recombinant Mullerian Inhibiting Substance on gynecological tumor cell lines and tumor explants. Cancer Res 49:2005–2011

34. Vescio RA, Connors KM, Youngkin T, Bordin GM, Robb JA, Umbreit JN, Hoffman RM (1990) Cancer biology for individualized cancer therapy: correlation of growth fraction index in native-state culture with tumor grade and stage. Proc Natl Acad Sci U S A 87:691–695

35. Vescio RA, Connors KM, Bordin GM, Robb JA, Youngkin T, Umbreit JN, Hoffman RM (1990) The distinction of small cell and non-small cell cancer by growth in native-state histoculture. Cancer Res 50:6095–6099

36. Li L, Hoffman RM (1991) Eye tissues grown in 3-dimensional histoculture for toxicological studies. J Cell Pharmacol 2:311–316

37. Hoffman RM (1991) Three-dimensional histoculture: origins and applications in cancer research. Cancer Cells 3:86–92

38. Hoffman RM (1991) *In vitro* sensitivity assays in cancer: a review, analysis and prognosis. J Clin Lab Anal 5:133–143

39. Vescio RA, Connors KM, Kubota T, Hoffman RM (1991) Correlation of histology and drug response of human tumors grown in native-state three-dimensional histoculture and in nude mice. Proc Natl Acad Sci U S A 88:5163–5166

40. Augustin HG, Koh GY (2017) Organotypic vasculature: From descriptive heterogeneity to functional pathophysiology. Science 357: pii: eaal2379

Chapter 10

Clinical Usefulness of the Histoculture Drug Response Assay for Breast Cancer

Robert M. Hoffman and Hirokazu Tanino

Abstract

The histoculture drug response assay (HDRA) was used to compare drug sensitivity of recurrent and primary breast cancer in vitro as well as in the clinic. The HDRA utilizes 3-dimensional culture of human tumors on Gelfoam®. The evaluation rate was 98.8%. The HDRA mean inhibition rate of primary tumors vs. recurrent tumors was, respectively, 57.9% and 38.6% for doxorubicin (DOX); 59.9% and 42.8% for mitomycin C (MMC); 49.0% and 33.4% for 5-fluorouracil (5-FU); and 34.5% and 16.0% for cisplatinum (CDDP). The recurrent cases were pretreated clinically with CAF (cyclophosphamide [CTX], DOX, and 5-FU), CEF (CTX, epirubicin [EPN], and 5-FU), or CMF (CTX, methotrexate [MTX], and 5-FU). 64.7% of the recurrent cases were resistant to all four agents tested compared to 27% of the primary cases. Only 5.9% of the recurrent cases were sensitive to three or more agents as opposed to 18% of the primary cases. The correlation of the HDRA results to clinical outcome in another breast-cancer study was 80.0% with 15 cases evaluated consisting of five true positives, three false positives, seven true negatives, and no false negatives. HDRA was also performed on surgical specimens of primary tumor and axillary lymph node metastasis from each of 30 breast cancer patients. The lymph-node metastases were more resistant than the primary tumor for DOX, 5-FU, and MMC, but not for CDDP. The data suggest that both primary tumor and metastases from individual patients should be tested in the HDRA to enhance clinical efficacy of chemotherapy. There also was a lack of correlation with breast cancer subtype and drug response in the HDRA, further suggesting the importance of individualized treatment for breast cancer patients afforded by the HDRA.

Key words Gelfoam® histoculture, Metastasis, HDRA, Breast cancer, Primary tumor

1 Introduction

Breast cancers are clinically heterogeneous and have different clinical courses and responses to chemotherapy even if they are of the same stage [1]. Breast cancer has been classified into four subtypes: luminal A, luminal B, HER-2 enriched, and triple-negative breast cancer (TNBC) depending on the presence or absence of estrogen/progesterone receptors (ER/RP) or HER-2 [2]. Currently clinical selection of chemotherapeutic agents usually depends on the size of the primary tumor, presence of axillary lymph node

Robert M. Hoffman (ed.), *3D Sponge-Matrix Histoculture: Methods and Protocols*, Methods in Molecular Biology, vol. 1760, https://doi.org/10.1007/978-1-4939-7745-1_10, © Springer Science+Business Media, LLC, part of Springer Nature 2018

metastasis, and distant metastasis, all of which greatly influence the prognosis of the patient. The therapeutic regimen overwhelmingly depends on physician's choice, rather than individualized scientific tests.

The histoculture drug response assay (HDRA) is an appropriate method for the culture of breast cancer since it allows the interstitial cells, which occupy the major portion of the tumor in breast cancer, to be cultured in their natural 3-dimnensional architecture along with the cancer cells [3, 4]. In the presents chapter, HDRA testing of primary and recurrent breast cancer in the HDRA is reviewed, demonstrating that recurrent breast cancer very frequently acquired multidrug resistance and that the HDRA is shown to be clinically useful for patients for recurrent breast cancer [5]. The HDRA results for primary and metastatic tumors of each patient was compared in another study.

Chemosensitivity assays for cancer can be divided into either 2-dimensional culture or 3-dimensional culture. Although 2-dimensional cell culture may be more convenient [6], it lacks tumor-stromal interaction within intact tumor tissue, which is more representative of the patient's tumor in vivo [6, 7]. The HDRA uses 3D Gelfoam® histoculture. The present chapter reviews the clinical usefulness of the HDRA for breast cancer [8].

2   Materials

1. Surgical knife.

2. Fine forceps.

3. Fine scissors.

4. Tissue and cell culture plate or flask: Tissue culture dishes (60 mm).

5. Culture medium (CM): RPMI1640 with L-glutamine and phenol red, 100 µM modified Eagle's medium (MEM) nonessential amino acids (Thermo Fisher Scientific), 1 mM sodium pyruvate, 50 µg/ml gentamycin sulfate, 2.5 µg/ml amphotericin B (Thermo Fisher Scientific), 15% fetal bovine serum (FBS). We recommend testing different batches of FBS before purchasing a large stock.

2.1 Gelfoam® Histoculture

1. Gelfoam® Absorbable Collagen Sponge USP, 12–7 mm format (Pfizer, NDC: 0009-031508) as an example.

2. Gelfoam® trimming aid: Double-edged blade.

3. Sterile flat weighing metal spatula.

4. Water-jacketed CO_2 incubator, set at 37 °C, 5% CO_2, ≥90% humidity.

5. Water bath.

6. Tissue transportation medium: Sterile phosphate buffer saline (PBS) or other solution with physiologic pH.

7. Sterile transportation container.

8. Ethanol solution (70%).

9. Disinfectant solution for biological waste disposal.

10. Collagenase type I (Sigma).

11. MTT (Sigma).

12. Sodium succinate (Wako Ind., Tokyo, Japan).

13. Dimethyl sulfoxide (DMSO) (Sigma).

14. Microplate reader (VersaMax, Sunnyvale, CA, USA).

2.2 Drugs

1. Doxorubicin (DOX).

2. Mitomycin C (MMC) .

3. 5-Fluorouracil (5-FU).

4. Cisplatinum (CDDP).

3 Methods

3.1 Patients (Wakayama Medical College)

Primary and recurrent breast tumors and lymph node metastasis were evaluated with the HDRA on tumor tissues obtained at surgery. The tumors were transported from Wakayama Medical College to the HDRA laboratory at the Eiken Co., Ltd., and kept at 4 °C in Hanks' solution with antibiotics until the next morning [5].

3.2 Patients (Kagoshima University Hospital)

The average age of the patients was 48.54 with a range from 24 to 87 years. The staging distribution was as follows: 18 patients (36%) had stage I; 17 patients (34%) had stage IIa; 8 patients (16%) had stage IIb; and 7 patients (14%) had stage IIIa. There were no patients with stage IIIb or over [9].

3.3 The Histoculture Drug Response Assay

1. Obtain tumor specimens from surgery.

2. Aseptically wash tumors in Hanks' balanced salt solution (HBSS).

3. Mince into fragments approximately 0.5 mm in diameter.

4. Exclude necrotic and nonviable portions using 3-(4,5 dimethylthiazol-2-yl)-5-(3-carboxymethoxyphenyl)-2-(4-sulfo-phenyl)-2-tetrazolium (MTS; Sigma, St. Louis, MO, USA) staining.

5. Weigh the viable cancer tissues on a chemical balance.

6. Immerse Gelfoam® in 1 ml culture medium in 24-well plates.

7. Place tumor fragments onto Gelfoam®.

8. Dissolve drugs at varying concentrations in cell culture medium.

9. Pipette the medium into each well (1 ml/well), taking care to avoid covering the Gelfoam® completely.

10. Incubate for 72 h at 37 °C with 5% CO_2.

11. Treat with 100 μl 0.06% collagenase type I (Sigma) in HBSS and 0.2% MTT (Sigma) in PBS containing 50 mM sodium succinate (Wako Ind., Tokyo, Japan).

12. Incubate plates for another 4 h, remove the medium, and add 0.5 mL dimethyl sulfoxide (DMSO) (Sigma) to each well to extract MTT formazan.

13. Transfer extracts from each well (100 μl) to a 96-well plate and measure the absorbance at 540 nm using a microplate reader.

14. Measure the weight of the residual tumor after extraction of MTT formazan and calculate the absorbance per g tumor.

15. Calculate the tumor inhibition rate (%), relative to the untreated control group using the following equation: tumor inhibition rate (%) = [1—(absorbance per g tumor in the treated group/absorbance per g tumor in the untreated group)] × 100. At each concentration, average the inhibition rates for replicate cultures to construct a dose-response curve.

3.4 Correlation of Chemosensitivity with the HDRA and Clinical Response

1. Categorize the HDRA results for each drug: drug resistant [IR greater than −0.5 standard deviation (SD) from the mean IR]; intermediate (IR between −0.5 SD and +0.5 SD from the mean IR); or sensitive (IR more than 0.5 + SD from the mean IR).

2. Calculate accurate prediction rate (%) = (no. of true-positive cases + no. of true-negative cases)/total no. of cases [10].

3.5 Statistical Analysis

1. Perform the χ^2 test and Fisher's exact test to compare the characteristics of patients and response rate using the Kaplan–Meier method for progression-free survival (PFS) and overall survival (OS) using SPSS ver.19.0 (SPSS Inc., Chicago, IL, USA) with a p-value less than 0.05 considered to be statistically significant [11].

4 Results

The sensitivity of recurrent breast cancer was significantly lower than that of primary breast cancer for all four agents tested in the HDRA with the response rates of recurrent tumors less than half that of primary tumors for three of the four agents tested. The mean inhibition rate of primary tumors vs. recurrent tumors in the HDRA was, respectively, 57.9% and 38.6% for DOX ($p < 0.0005$);

Table 1

Positive response rate of primary and recurrent breast cancer [5]

Drugs	Cutoff Concentration (μg/ml)	Inhibition rate (%)	Positive response rate (%)		Clinical Efficacy rate (%)
			Primary	Recurrent	
DOX	15	60	61.0	33.3	35
5-FU	300	60	34.5	15.0	26
MMC	2.0	70	47.1	15.6	38
CDDP	20	40	43.4	18.2	40

59.9% and 42.8% for MMC ($p < 0.01$); 49.0% and 33.4% for 5-FU ($p < 0.01$); and 34.5% and 16.0% for CDDP ($p < 0.005$), respectively [5] (Table 1) (*see* **Note 1**).

The recurrent cases were treated clinically with CAF (CTX, DOX, and 5-FU), CEF (cyclophosphamide, epirubicin, and 5-FU), or CMF (cyclophosphamide, methotrexate, and 5-FU). In the CAF and CEF groups, the HDRA sensitivity to CDDP was significantly lower in recurrent disease than that of primary breast cancer ($p < 0.005$). 64.7% of the recurrent cases were resistant to all four agents tested as opposed to 27.3% of the primary cases and only 5.9% of the recurrent cases were sensitive to three or more agents, as opposed to 18.2% of the primary cases [5] (Table 2) (*see* **Note 2**).

The correlation of the HDRA chemosensitivity results to clinical outcome in another study was 80.0% with 14 recurrent cases (15 tumors) evaluated with 5 true positives, 3 false positives, 7 true negatives, and no false negatives [5] (Table 3).

The HDRA chemosensitivity of the primary tumors was higher than that of the lymph node metastases with a statistically-significant difference at $p < 0.05$ for DOX, 5-FU, and MMC but not for CDDP [4] (*see* **Notes 3** and **4**).

Paclitaxel (PTX) sensitivity in the HDRA was an independent prognostic predictor for disease-free survival (DFS) (relative risk (RR) = 0.0097; 95% confidence interval (CI) = 0.0038–0.87; $p = 0.003$c) (Fig. 1) [9].

In another study to determine whether breast cancer subtype could be distinguished in the HDRA, immunohistochemical staining identified 37 ER-positive patients (74%), 32 PR-positive patients (64%), 15 HER-2-positive patients (30%), and 8 TNBC patients (16%). Based on the immunohistochemical study results, the breast cancer patients were classified into four subtypes. Luminal A type was most frequent with 27 patients (54%) [8].

Table 2

HDRA results of the primary and recurrent tumors tested against all four agents [5]

Number of positive agents	Primary	Recurrent
4	13 (13.1)%	0 (0.0)%
3	18 (18.2)%	1 (5.9)%
2	24 (24.2)%	2 (11.8)%
1	17 (17.2)%	3 (17.6)%
0	27 (27.3)%	11 (64.7)%
Total	99 (100.0)	17 (100.0)

Table 3

Correlation between the results of HDRA and clinical response of patients with recurrent breast cancer [5]

Correlation	Drugs DOX	5-FU	MMC	CDDP	Chemotherapy	Clinical response	Site of measurable disease
TP	73.4	56.9	67.8	21.5	FAM	CR	Lung
TP	64.1	35.0	72.1	31.9	MMC + CAP + EPN	PR	Liver
TP	70.1	35.4	65.6	9.0	CAF	PR	Skin
TP	40.0	70.5	81.3	67.0	CAF	PR	Lung
TP	78.2	63.4	76.1	34.8	CAF	PR	LN
FP	63.0	64.8	61.9		5-FU + CDDP	NC	LN
FP	70.7	28.0	71.2	56.3	CAF	NC	Skin
FP	66.6	46.3	75.6	25.4	MMC	PD	Skin
TN				17.2	5-FU + CDDP	NC	Liver
TN	58.4	55.7	53.7	28.2	CDDP + DOX	NC	Liver
TN	35.5	26.1	50.1	15.3	FP	PD	Skin
TN	27.3	54.5	57.8	21.4	CMF	PD	LN
TN	63.0	64.8	61.9		MMC	PD	LN
TN	25.3			42.6	CAF	PD	Lung
TN	66.9	11.2	65.1	0.0	5-FU	PD	Lung

TP: true positive; *FP*: false positive; *TN*: true negative; *Underline*: Chemosensitive estimated by HDRA; *ND* not done, *MCC* mitomycin C, *CPA* cyclophosphamide, *EPN* epirubicin, *DOX* doxorubicin, *5-FU* 5-fluorourascil, *MTX* methotrexate, *CDDP* cisplatinum, *CAF* CTX + DOX + 5-FU, *CMF* CTX + MTX + 5-FU

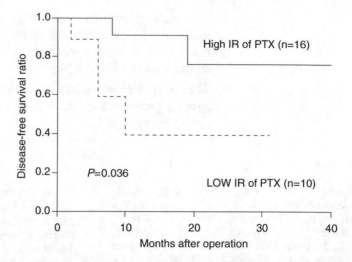

Fig. 1 Kaplan-Meier disease-free survival curves of patients with breast cancer whose tumors showed high vs. low inhibition rates associated with paclitaxel [9]

After the subtype was determined for each patient, the HDRA was conducted and the average of the IR of each chemotherapeutic agent was calculated for each subtype [8].

One-way ANOVA test demonstrated no significant correlation between IR of any agent or combination with breast cancer subtype ($p = 0.851$) [8] (*see* **Note 5**).

5 Notes

1. Cutoff concentrations and cutoff IRs were modified to fit historical response rates for each drug in breast cancer [2]. For DOX the cutoff concentration was determined to be 15 µg/ml with a cutoff IR of 60%. For 5-FU, the cutoff concentration was determined to be 300 µg/ml with a cutoff IR of 60%. For MMC, the cutoff concentration was determined to be 2.0 µg/ml with a cutoff IR of 70%. For CDDP, the cutoff concentration was determined to be 20 µg/ml with a cutoff IR of 50% [5].

2. 64.7% of the recurrent cases were resistant to all four drugs tested as opposed to only 27.3% of the primary cases. Only one recurrent cases was sensitive to three or more drugs as opposed to 18 of the primary cases. Recurrent breast cancer requires the HDRA to be able to identify effective drugs.

3. It is well known that the malignant tumor consists of heterogeneous clones of cancer cells that metastasize to other organs [12]. This suggests that the drug sensitivity of metastatic lesions might be different from that of primary lesion [4].

4. The data presented here suggest that both primary and metastatic lesions should thus be tested, whenever possible, in the HDRA to ensure maximum clinical efficacy of drugs found effective in the HDRA [4].

5. The study demonstrates that the efficacy of a chemotherapeutic agent depends on the individual tumor rather than the breast cancer subtype [8].

References

1. Sorlie T, Perou CM, Tibshirani R, Aas T, Geisler S, Johnsen H, Hastie T, Eisen MB, Van de Rijin M, Jeffrey SS, Thorsen T, Quist H, Matese JC, Brown PO, Botstein D, Lonning PE, Borresen-Dale AL (2001) Gene expression pattern of breast carcinomas distinguish tumor subclasses with clinical implications. Proc Natl Acad Sci U S A 98:10869–10874

2. Robson M, Im SA, Senkus E, Xu B, Domchek SM, Masuda N, Delaloge S, Li W, Tung N, Armstrong A, Wu W, Goessl C, Runswick S, Conte P (2017) Olaparib for metastatic breast cancer in patients with a germline BRCA mutation. N Engl J Med 377:523–533

3. Hoffman, R.M. Histocultures and their use. In: Encyclopedia of Life Sciences. John Wiley and Sons, Ltd.: Chichester, 2010, Published Online. DOI: 10.1002/9780470015902. a0002573.pub2

4. Furukawa T, Kubota T, Tanino H, Oura S, Yuasa S, Murate H, Morita K, Kozakai K, Yano T, Hoffman RM (2000) Chemosensitivity of breast cancer lymph node metastasis compared to the primary tumor from individual patients tested in the histoculture drug response assay. Anticancer Res 20:3657–3658

5. Tanino H, Oura S, Hoffman RM, Kubota T, Furukawa T, Arimoto J, Yoshimasu T, Hirai I, Bessho T, Suzuma T, Sakurai T, Naito Y (2001) Acquisition of multidrug resistance in recurrent breast cancer demonstrated by the histoculture drug response assay. Anticancer Res 21:4083–4086

6. Kim JB, Stein R, O' Hare MJ (2004) Three-dimensional *in vitro* tissue culture models of

breast cancer- a review. Breast Cancer Res Treat 85:281–291

7. Hoffman RM (1993) *In vitro* assays for chemotherapy sensitivity. Crit Rev Oncol Hematol 15:99–111

8. Kim KY, Chung BW, Yang I, Kim MB, Hoffman RM (2014) Independence of cytotoxic drug sensitivity profiles and receptor subtype of invasive ductal breast carcinoma demonstrated by the histoculture drug response assay (HDRA). Anticancer Res 34:7197–7201

9. Shinden Y, Kijima Y, Hirata M, Arima H, Nakajyo A, Tanoue K, Maemura K, Natsugoe S (2016) Clinical significance of the histoculture drug response assay in breast cancer. Anticancer Res 36:6173–6178

10. Nakada S, Aoki D, Ohie S, Horiuchi M, Suzuki N, Kanasugi M, Susumu N, Udagawa Y, Nozawa S (2005) Chemosensitivity testing of ovarian cancer using the histoculture drug response assay: sensitivity to cisplatin and clinical response. Int J Gynecol Cancer 15:445–452

11. Jung PS, Kim DY, Kim MB, Lee SW, Kim JH, Kim YM, Kim YT, Hoffman RM, Nam JH (2013) Progression-free survival is accurately predicted in patients treated with chemotherapy for epithelial ovarian cancer by the histoculture drug response assay in a prospective correlative clinical trial at a single institution. Anticancer Res 33:1029–1034

12. Bhatia S, Frangioni JV, Hoffman RM, Iafrate AJ, Polyak K (2012) The challenges posed by cancer heterogeneity. Nat Biotechnol 30:604–610

Chapter 11

Clinical Usefulness of the Histoculture Drug Response Assay for Prostate Cancer and Benign Prostate Hypertrophy (BPH)

Robert M. Hoffman

Abstract

The histoculture drug response assay (HDRA) has been adapted to determine androgen sensitivity in Gelfoam histoculture of human benign prostatic tissue as well as prostate cancer. Gelfoam histoculture was used to measure androgen-independent and androgen-dependent growth of benign and malignant prostate tissue. The androgen-sensitivity index was significantly higher in 23 paired specimens of prostate cancer compared to benign prostate hypertrophy (BPH). Genistein decreased the androgen-sensitivity index of BPH and prostate cancer in Gelfoam® histoculture in a dose-dependent manner.

Key words Gelfoam® histoculture, Prostate cancer, Benign prostate hypertrophy (BPH), Dihydrotestosterone (DHT), Hydroxyflutamide (HF), DHT/HF index, Androgen-sensitivity index

1 Introduction

Prostate cancer is epidemic in the Western world. Benign prostatic hypertrophy (BPH) is also a very common disorder in the aging population in the Western world [1].

The histoculture drug response assay (HDRA) on Gelfoam® [2–10] provides a realistic model of in vivo tissue growth in vitro, since stroma and epithelium are maintained in anatomical juxtaposition in their natural geometry on the flexible Gelfoam®. The growth medium can also be carefully defined.

The HDRA was used to establish androgen sensitivity in BPH prostate and normal nonprostate tissues using the ratio of [³H] thymidine incorporation in dihydrotestosterone (DHT)-treated compared to hydroxyflutamide (HF)-treated histocultures. HF is an anti-androgen whose mechanism of action is to block androgen receptor binding to DHT and thereby to effectively block androgen-mediated action. This ratio is the androgen sensitivity index [2]. The androgen sensitivity index was compared in BPH and prostate cancer [11].

Robert M. Hoffman (ed.), *3D Sponge-Matrix Histoculture: Methods and Protocols*, Methods in Molecular Biology, vol. 1760, https://doi.org/10.1007/978-1-4939-7745-1_11, © Springer Science+Business Media, LLC, part of Springer Nature 2018

Genistein, which is a major ingredient of bean curd (tofu), which in turn is made from the soybean, is an isoflavonoid [12, 13]. Peterson and Barnes [14] have shown that genistein decreases the growth of LNCaP, a hormone-dependent human prostate cancer cell line, as well as DU-145, a hormone-independent human prostate cancer cell line [15]. The efficacy of genistein on the androgen sensitivity index in both BPH and malignant prostate cancer was also determined in Gelfoam® histoculture [15].

2   Materials

1. Fine forceps.

2. Fine scissors.

3. Tissue plate.

4. Culture medium (CM): RPMI1640 with L-glutamine and phenol red, 100 μM modified Eagle's medium (MEM) nonessential amino acids (Thermo Fisher Scientific), 1 mM sodium pyruvate, 50 μg/mL gentamycin sulfate, 2.5 μg/mL amphotericin B (Thermo Fisher Scientific), 15% fetal bovine serum (FBS). We recommend testing different batches of FBS before purchasing a large stock.

5. Gelfoam® Absorbable Collagen Sponge USP, 12–7 mm format (Pfizer, NDC: 0009-031508) as an example.

6. Gelfoam® trimming aid: Double-edged blade.

7. Sterile flat weighing metal spatula.

8. Water-jacketed CO_2 incubator, set at 37 °C, 5% CO_2, ≥90% humidity.

9. Tissue transportation medium: Sterile phosphate buffered saline (PBS) or other solution with physiologic pH.

10. Sterile transport container.

11. Sterile scalpels and blades.

12. Disinfectant solution for biological waste disposal.

13. Ethanol solution (70%).

14. Water bath.

15. Tissue transportation medium: PBS or other solution with physiologic pH.

16. Sterile transportation container.

17. Disinfectant solution for biological waste disposal.

18. [^{3}H]thymidine.

19. Tekmar homogenizer (Janke & Kunkel, Staufen, Germany).

20. Hyamine

3 Methods

3.1 Tissue Preparation

1. Place specimens from patients with early prostate cancer or BPH specimens in Eagle's minimal essential medium (MEM) solution on ice in the operating room for transport.

2. Bring the specimens to a pathologist to weigh and measure the specimens and mark the prostate surface with green ink to aid in the pathological staging of the tumor.

3. Cut the prostate tissue serially as indicated below.

4. Indicate the anterior and posterior parts of the prostate tissue.

5. Make a sagittal cut running from distal to proximal, dividing the prostate into right and left sections.

6. Make approximately 0.5 cm cuts beginning at the distal urethral end of the prostate.

7. Identify each section with a letter label.

8. Select representative specimens of cancer and benign prostate hypertrophy (BPH) grossly for Gelfoam® histoculture.

9. Keep anatomical mirror-image sections of the specimens for Gelfoam® histoculture and for microscopic anatomy.

10. Send to the laboratory in ice-cold culture solution [11].

3.2 Histoculture

1. Cut the prostate tissue into small 1.0 mm^3 pieces and then thoroughly mix.

2. Plant these tissue minces onto previously hydrated Gelfoam®, in 6-well plates.

3. For each tissue studied, treat two plates with DHT alone (treated plates).

4. Choose a DHT concentration of 2×10^{-8} M as a near-maximal stimulus for growth stimulation [16].

5. Add DHT (2×10^{-8} M) plus HF (2×10^{-3} M) to control plates to block androgen [2] (*see* **Note 1**).

6. Add DHT in 5 μl ethanol daily to a final concentration of 2×10^{-8} M in each well.

7. Add HF plus DHT in 5 μl ethanol daily to the control plates.

8. Add 8 μC of [^{3}H]thymidine to each well and stop the experiment 24 h later.

3.3 Processing Tissue for [^{3}H] thymidine Uptake and Protein Assay

1. Pool tissue explants from each plate and wash until an aliquot of 1/20 of the saline wash is less than 100 cpm.

2. Place the tissue in a test tube in a beaker of ice and homogenize in 0.05 M Tris buffer, pH 7.8, using a Tekmar homogenizer operating at 3000 rpm for 10-s bursts times 6.

3. Centrifugate the homogenate in a refrigerated centrifuge at 3000 rpm, and remove the supernatant for protein assay.

4. Digest the precipitate, which contains DNA, in 0.3 ml hyamine at 70–80 °C for 2 h.

5. Count the solution for [^{3}H]thymidine CPM in a liquid scintillation counter with an efficiency of about 60% for tritium [2].

6. Assay the supernatant for protein by the method of de Boer [17] or equivalent.

7. Express androgen sensitivity as the ratio of [^{3}H]thymidine/μg protein in the DHT histoculture/[^{3}H]thymidine/μg protein in the DHT + HF histoculture, since HF at 2×10^{-5} M effectively blocks DHT receptor binding [2, 18] (*see* **Notes 2** and **3**).

4   Results

4.1   Androgen Sensitivity Index of BPH in Gelfoam® Histoculture

The ratio of [3H]thymidine/μg protein in DHT-treated BPH histocultures [^{3}H]thymidine/μg protein in control histocultures (HF + DHT) provides a DHT/HF ratio, that is, the androgen sensitivity index (*see* **Notes 4**) [2].

4.2   Androgen Stimulation of [³H]thymidine Incorporation in Gelfoam® Histocultures of Human BPH

DHT (2×10^{-8} M), added daily to BPH histocultures increased [^{3}H]thymidine/μg protein an average of 2.2 times in comparison to HF + DHT controls. A paired t-test showed a significant stimulation of [^{3}H]thymidine uptake ($P = 0.0001$) in the DHT-treated samples for 19 prostates studied [2].

4.3   Androgen Sensitivity Index in BPH Vs. Prostate Cancer in Gelfoam® Histoculture

[^{3}H]thymidine incorporation was higher in BPH histocultures compared to prostate cancer histocultures incubated with DHT alone. However, the androgen sensitivity index was greater in prostate cancer specimens compared to BPH from the same prostate ($P \leq 0.03$) in Gelfoam® histoculture. There was no correlation of the androgen sensitivity index to Gleason grade of the prostate cancer [11].

4.4   Genistein Inhibits [³H]thymidine Incorporation in BPH and Prostate Cancer

Genistein significantly inhibited [^{3}H]thymidine incorporation/μg protein in BPH in Gelfoam® histoculture in a dose-response fashion from 1.25 μg/mL to 10 μg/ml (46 μM). Genistein efficacy on [^{3}H] thymidine incorporation in prostate cancer Gelfoam® histoculture was similar to those noted in BPH Gelfoam® histoculture at similar concentrations of genistein. The fact that BPH and cancer from the same prostate in the same experiments showed similar responses to genistein suggests that the effects in many, if not all, of the cancers would be similar to those noted in BPH. However, genistein did not affect cell viability [15] (Figs. 1 and 2) (*see* **Note 5**).

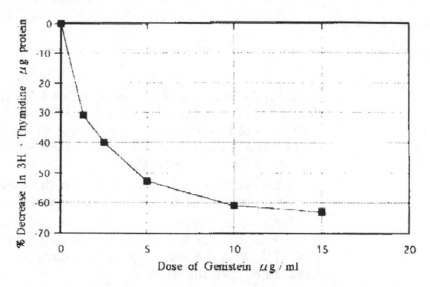

Fig. 1 Effect of genistein on BPH in Gelfoam® histoculture, demonstrating a decrease in [³H]thymidine incorporation per µg protein in human BPH tissue in Gelfoam® histoculture with varying concentrations of genistein, beginning at 1.25 µg/ml and extending to 15 µg/ml. Note the rather sharp decline in [³H]thymidine incorporation occurring in the range of 1.25–10 µg/ml, with relative leveling off thereafter [15]

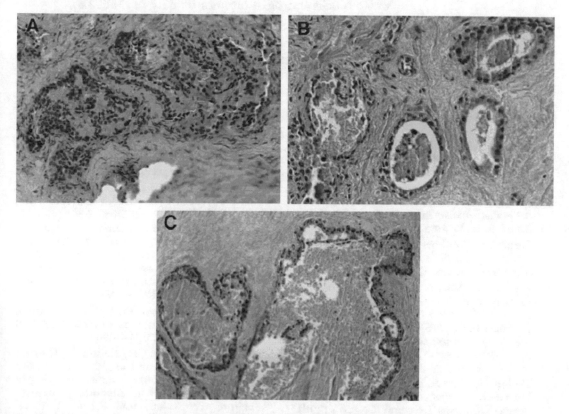

Fig. 2 Histological effects of DHT and genistein on BPH in Gelfoam® histoculture (**a**) Microscopic appearance of BPH tissue prior to histoculture (H&E stain, ×200). (**b**) Microscopic appearance of BPH tissue following 5 days of histoculture with DHT at 2×10^{-8} M, added each day from days 2 to 5. Note mild epithelial cell nuclear pleomorphism and sloughing of debris into acini (H&E stain, ×200). (**c**) Microscopic appearance of BPH tissue following 5 days of histoculture with genistein (5 µg/mL), added on day 2, and DHT (2×10^{-8} M), added daily from days 2 to 5. Note mild nuclear pleomorphism with sloughing of debris into acini; also note mild variable decreases in stromal cell nuclei (H&E stain, ×100) [15]

5 Notes

1. For optimal stimulation of prostate growth, a DHT concentration of 2×10^{-8} M was chosen, since it is close to the physiologic levels of DHT found in the human prostate. Shao et al. [16] showed this to be an optimal concentration for stimulation of growth of the rat prostate [11].

2. Kennealey and Furr [18] showed that HF at this concentration effectively blocks DHT binding to the androgen receptor [2].

3. HF blockade interrupts the pathway of androgen-mediated action and prevents any biological androgen response. The DHT/HF ratio, therefore, appears to describe the androgen sensitivity of the tissue tested.

4. This study established a simple technique for assaying androgen sensitivity in Gelfoam® histoculture by measuring the DHT + HF ratio for [³H]thymidine uptake per microgram of protein. This assay is based on a highly specific effect of HF, which blocks androgen receptor binding of DHT [2].

5. Genistein decreases the growth of both BPH and prostate cancer tissue in histoculture. The data suggest that genistein has the potential as a therapeutic agent for BPH and prostate cancer [15].

References

1. Logothetis CJ (2017) Improved outcomes in men with advanced prostate cancer. N Engl J Med 377:388–390

2. Geller J, Sionit LR, Connors K, Hoffman RM (1992) Measurement of androgen sensitivity in the human prostate in *in vitro* three-dimensional histoculture. Prostate 21:269–278

3. Furukawa T, Kubota T, Hoffman RM (1995) Clinical applications of the histoculture drug response assay. Clin Cancer Res 1:305–311

4. Kubota T, Sasano N, Abe O, Nakao I, Kawamura E, Saito T, Endo M, Kimura K, Demura H, Sasano H, Nagura H, Ogawa N, Hoffman RM (1995) Potential of the histoculture drug response assay to contribute to cancer patient survival. Clin Cancer Res 1:1537–1543

5. Tanino H, Oura S, Hoffman RM, Kubota T, Furukawa T, Arimoto J, Yoshimasu T, Hirai I, Bessho T, Suzuma T, Sakurai T, Naito Y (2001) Acquisition of multidrug resistance in recurrent breast cancer demonstrated by the histoculture drug response assay. Anticancer Res 21:4083–4086

6. Singh B, Li R, Xu L, Poluri A, Patel S, Shaha AR, Pfister D, Sherman E, Hoffman RM, Shah J (2002) Prediction of survival in patients with head and neck cancer using the histoculture drug response assay. Head Neck 24:437–442

7. Jung PS, Kim DY, Kim MB, Lee SW, Kim JH, Kim YM, Kim YT, Hoffman RM, Nam JH (2013) Progression-free survival is accurately predicted in patients treated with chemotherapy for epithelial ovarian cancer by the histoculture drug response assay in a prospective correlative clinical trial at a single institution. Anticancer Res 33:1029–1034

8. Freeman A, Hoffman RM (1986) *In vivo*-like growth of human tumors *in vitro*. Proc Natl Acad Sci U S A 83:2694–2698

9. Vescio RA, Redfern CH, Nelson TJ, Ugoretz S, Stern PH, Hoffman RM (1987) *In vivo*-like drug response of human tumors growing in three-dimensional, gel-supported, primary culture. Proc Natl Acad Sci U S A 84:5029–5033

10. Ji WB, Um JW, Ryu JS, Hong KD, Kim JS, Min BW, Joung SY, Lee JH, Kim YS (2017) Clinical significance of 5-fluorouracil chemosensitivity testing in patients with colorectal cancer. Anticancer Res 37:2679–2682

11. Geller J, Partido C, Sionit L, Youngkin T, Nachtsheim D, Espanol M, Tan Y, Hoffman RM (1997) Comparison of androgen-independent growth and androgen-dependent growth in BPH and cancer tissue from the same radical prostatectomies in sponge-gel matrix histoculture. Prostate 31:250–254

12. Adlercreutz H, Markkanen H, Watanabe S (1993) Plasma concentrations of phyto-estrogens in Japanese men. Lancet 342:1209–1210

13. Pavese JM, Krishna SN, Bergan RC (2014) Genistein inhibits human prostate cancer cell detachment, invasion, and metastasis. Am J Clin Nutr 100(Suppl 1):431S–436S

14. Peterson G, Barnes SL (1993) Genistein and biochanin A inhibit the growth of human prostate cancer cells but not epidermal growth factor receptor tyrosine autophosphorylation. Prostate 22:335–345

15. Geller J, Sionit L, Partido C, Li L, Tan X-Y, Youngkin T, Nachtsheim D, Hoffman RM (1998) Genistein inhibits the growth of human-patient BPH and prostate cancer in histoculture. Prostate 34:75–79

16. Shao TC, Tindall DJ, Cunningham GR (1986) Age dependency of androgen and estrogen effects on incorporation of (3H)thymidine by rat prostates in organ culture. Prostate 8:349–362

17. de Boer W, Bolt J, Kuiper GG, Brinkmann AO, Mulder E (1987) Analysis of steroid- and DNA-binding domains of the calf uterine androgen receptor by limited proteolysis. J Ster Biochem 28:9–19

18. Kennealey G, Furr B (1991) Use of the nonsteroidal anti-androgen casodex in advanced prostatic carcinoma. Urol Clin North Am 18:99–110

In Vivo-Like Cell-Cycle Phase Distribution of Cancer Cells in Gelfoam® Histoculture Observed in Real Time by FUCCI Imaging

Robert M. Hoffman and Shuya Yano

Abstract

FUCCI color codes cells as they express different color fluorescent proteins as they go through phases of the cell cycle. The cell cycle phase distribution of cancer cells in Gelfoam® histoculture was similar to in vivo tumors, whereby only the surface cells proliferate and interior cells are quiescent in G_0/G_1. In contrast, in 2D cancer-cell culture, most of the cells are always cycling. The cancer cells responded similarly to toxic chemotherapy in Gelfoam® culture as in vivo; as such, therapy is cell cycle dependent. In 2D culture, cancer cells were much more chemosensitive. These results indicate why the drug response pattern of tumors in Gelfoam® histoculture reflects what is observed in vivo.

Key words FUCCI (fluorescence ubiquitination cell cycle indicator), Color-coding, Gelfoam® histoculture, 3D culture, Imaging, Cell cycle, Chemotherapy

1 Introduction

Sakaue-Sawano et al. [1] have demonstrated that the cell-cycle phase can be visualized by color-coded imaging using a fluorescent ubiquitination-based cell cycle indicator (FUCCI). In the FUCCI system, G_0/G_1 cells express a red-orange fluorescent protein and express a green fluorescent protein in S/G_2. Intravital FUCCI imaging of cell-cycle dynamics of cancer cells has demonstrated that 90% of cancer cells in the center and 80% of total cells of an established tumor are in G_0/G_1 phase [2]. FUCCI imaging demonstrated that cytotoxic agents killed only proliferating cancer cells at the surface and, in contrast, had little effect on quiescent cancer cells, which are the vast majority of an established tumor [2]. Resistant quiescent cancer cells restarted cycling after the cessation of chemotherapy [2]. The cell-cycle phase distribution is a major factor in chemoresistance. We review in this chapter that cancer cells in Gelfoam histoculture have an in vivo-like cell cycle pattern,

Robert M. Hoffman (ed.), *3D Sponge-Matrix Histoculture: Methods and Protocols*, Methods in Molecular Biology, vol. 1760, https://doi.org/10.1007/978-1-4939-7745-1_12, © Springer Science+Business Media, LLC, part of Springer Nature 2018

demonstrated by FUCCI imaging, unlike cancer cells in 2D histoculture, and can therefore be used for discovery of novel effective therapeutics.

2 Materials

1. MKN45, a poorly differentiated stomach adenocarcinoma-derived cells from a liver metastasis of a patient [3].
2. RPMI 1640 medium with 10% fetal bovine serum and penicillin/streptomycin [2–4].
3. Plasmids expressing mKO2-hCdt1 (orange fluorescent protein) or mAG-hGem (green fluorescent protein) (Medical & Biological Laboratory, Nagoya, Japan).
4. Lipofectamine™ LTX (Invitrogen, Carlsbad, CA).
5. FACSAria cell sorter (Becton Dickinson, Franklin Lakes, NJ).

2.1 Gelfoam® Histoculture Materials

1. Gelfoam® Absorbable Collagen Sponge USP, 12–7 mm format (Pfizer, NDC: 0009-0315-08) (Gelfoam®) as an example.
2. Sterile Petri dishes, 100 mm × 20 mm or wider or 6-well tissue culture plates.
3. Phosphate-buffered saline (PBS) [2–4]
4. Sterile metal forceps or tweezers.
5. Sterile metal scissors.
6. Sterile flat weighing metal spatula.
7. Water-jacketed CO_2 incubator, set at 37 °C, 5% CO_2, ≥90% humidity.
8. Water bath.

2.2 Confocal Fluorescence Microscopy

1. FV1000 confocal microscope (Olympus, Tokyo, Japan) or equivalent.

3 Methods

3.1 Establishment of MKN45 Cells Stably Transfected with FUCCI-Vector Plasmids

1. Use the FUCCI (fluorescent ubiquitination-based cell cycle indicator) expression system for cell cycle phase visualization.
2. Transfect plasmids expressing orange-red mKO2-hCdt1 into MKN45 cells using Lipofectamine™ LTX (Invitrogen, Carlsbad, CA).
3. Incubate the cells for 48 h after transfection and then trypsinize and seed in 96-well plates at a density of 10 cells/well.

4. Sort cells for orange-red using a FACSAria cell sorter (Becton Dickinson, Franklin Lakes, NJ).

5. Re-transfect the first-step-sorted orange-red fluorescent cells with the green mAG-hGem and then sort by green fluorescence [1–3, 5].

3.2 Gelfoam® Histoculture

1. Cut the prepared sterile Gelfoam® sponges to 1 cm cubes.

2. Place the Gelfoam® cubes in 6-well tissue culture plates.

3. Add RPMI 1640 medium to the Gelfoam® and incubate at 37 °C in order that the Gelfoam® absorbs the medium.

4. Seed cancer cells (1×10^6) expressing FUCCI on top of the hydrated Gelfoam® and incubate for 1 h.

5. Carefully add medium to the top of the Gelfoam®.

6. Incubate the Gelfoam® cell cultures at 37 °C in a humidified incubator with 5% CO_2 [3, 5–10].

3.3 Confocal Microscopy

1. Perform FUCCI imaging of cells in Gelfoam® histoculture with a confocal fluorescence microscope.

2. Import the tracing data to Volocity 6.0 version (Perkin Elmer, Waltham, MA), where all further analysis can be performed [2, 3].

4 Results and Discussion

4.1 Gelfoam® Histoculture of Cancer Cells

FUCCI-expressing MKN45 cells formed structures similar to tumors after seeding in Gelfoam® histoculture. The cancer cells forming tumors on Gelfoam® brightly expressed either mK02-hCdt1 (orange fluorescence) in G_0/G_1 or mAG-hGem (green fluorescence) in $S/G_2/M$ (Fig. 1) (*see* **Note 1**) [3].

4.2 Comparison of Cell Cycle Phase Distribution of FUCCI-Expressing Cells Cultured in Monolayer, Sphere, Gelfoam®, and In Vivo

In monolayer culture, in both the central and edge areas, approximately 50% of the cells were in $S/G_2/M$. In tumor spheres, most of the cells were in G_0/G_1 at both the surface and center. In both tumors and in Gelfoam® histoculture, the majority of the surface cells of the tumor were in $S/G_2/M$. In contrast, in the central area of the tumor, only approximately 10% of the cells were in $S/G_2/M$ (Fig. 2). A comparison was made of the cell cycle phase distribution in a subcutaneous tumor, liver tumor, and Gelfoam®, all formed from FUCCI-expressing MKN45 stomach cancer cells. At the early stages of each tumor, whether subcutaneous or in the liver, or on Gelfoam®, approximately 90% of the cells were in $S/G_2/M$. In contrast as each tumor matured, approximately 80% of the cells were in G_0/G_1. The early-stage and mature-stage cell cycle phase distribution was very similar for tumors in vivo and structures in Gelfoam® (Fig. 2) (*see* **Note 2**) [3].

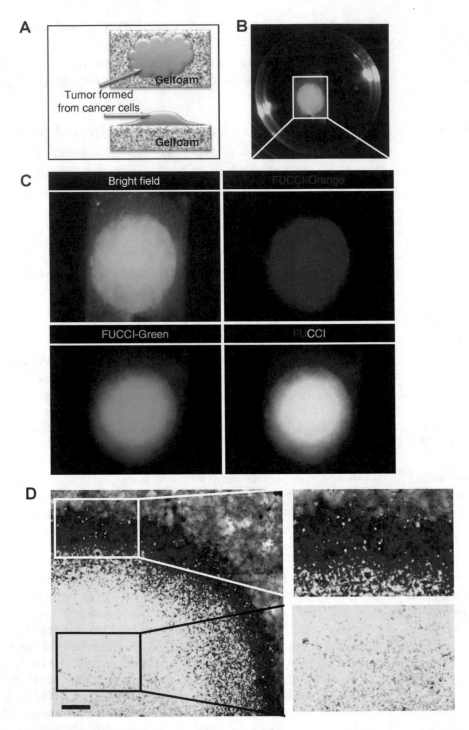

Fig. 1 Gelfoam® histoculture of FUCCI-expressing cancer cells. (**a**) Schema of FUCCI-expressing MKN45 stomach cancer cells forming a tumor on Gelfoam®. (**b**) Macroscopic appearance of the tumor formed on Gelfoam® histoculture. (**c**) Macro images of a tumor formed on Gelfoam® demonstrating FUCCI fluorescence. (**d**) FUCCI-expressing cancer cells in the tumor formed on Gelfoam®. Images at the single-cell level were acquired by confocal laser scanning microscopy. High-magnification images (×10) of an invading area of the tumor (upper right) and a non-invading area (lower right) of the tumor on Gelfoam® [3]

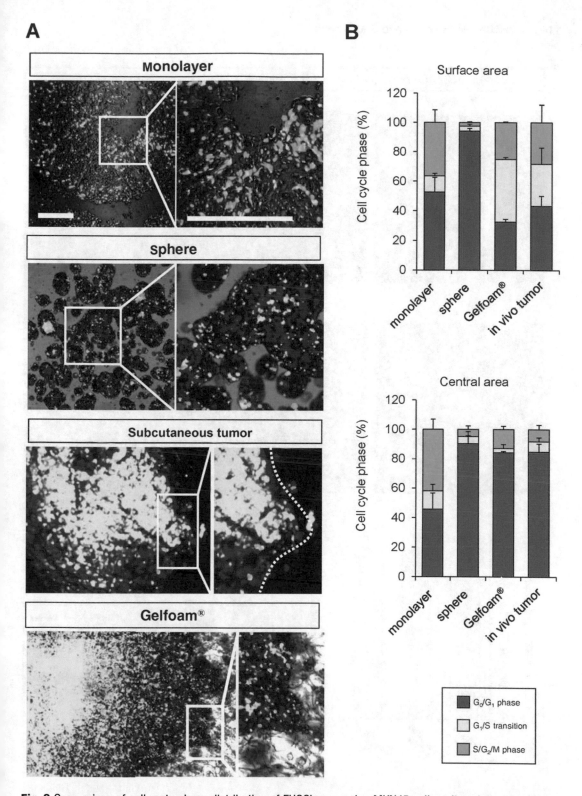

Fig. 2 Comparison of cell cycle phase distribution of FUCCI-expressing MKN45 cells cultured as monolayers, and on Gelfoam® and in vivo. (**a**) Representative images of FUCCI-expressing MKN45 cells cultured as monolayers, as spheres on agar and Gelfoam®, and in vivo. (**b**) Histograms show the cell-cycle phase distribution in the central area and invading area of the cultures and in vivo. Cancer cells in monolayer culture divide continuously. Cancer cells on agar aggregate and mostly remain in G_0/G_1 phase. Cancer cells on and in Gelfoam® have an in vivo-like cell cycle distribution.

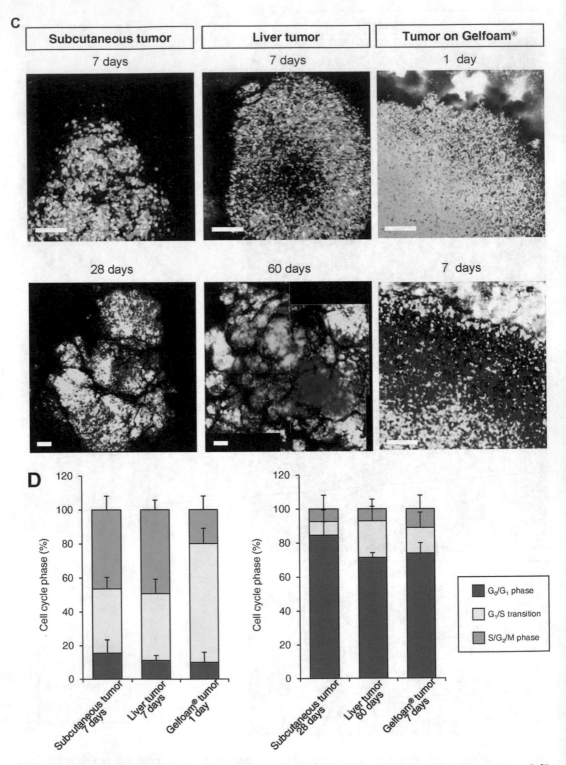

Fig. 2 (continued) (**c**) Representative time-course images of FUCCI-expressing subcutaneous tumor (left), FUCCI-expressing liver tumor (middle), and tumor formed from FUCCI-expressing cancer cells on Gelfoam® (right) at early and later stages. (**d**) Histograms show the cell-cycle phase distribution in early-stage and late-stage tumors growing subcutaneously, or in the liver or on Gelfoam®. Scale bars: 500 mm [3]

4.3 Cancer Cells in Gelfoam® Histoculture and In Vivo Tumors Have Similar 3-Dimensional Spatial Cell Cycle Phase Distribution

In both tumors in vivo and tumor-like structures in Gelfoam® culture, cancer cells were proliferating only near the surface of the tumor. At deeper levels, the vast majority of the cells were in G_0/G_1 in both tumors and on tumor-like structures in Gelfoam® (Fig. 3) [3].

4.4 Cancer Cells on Gelfoam®, but Not 2D Culture, Have the Same Cell Cycle Response to Cytotoxic Agents as In Vivo Tumors

Most chemotherapy agents target only proliferating cells and have little effect on quiescent cells. In monolayer culture, chemotherapy blocked cancer cells in G_2/M phase resulting in cell death. In sphere culture, chemotherapy had little effect since most cancer cells were in G_0/G_1, where they remained even after chemotherapy. In Gelfoam® histoculture and in subcutaneous tumors, chemotherapy targeted only proliferating cancer cells and had little effect on quiescent cancer cells, which were the majority of the cells. In both the subcutaneous tumors and the tumors in Gelfoam® histoculture, chemotherapy killed the surface proliferating cells, but the remaining cells were blocked in G_0/G_1 and resistant to chemotherapy as observed by FUCCI imaging (Fig. 4) (*see* **Note 3**) [3].

4.5 Cancer Cells on Gelfoam® Have Similar Spatial-Temporal Recovery from Cisplatinum and Paclitaxel Treatment as In Vivo Tumors, in Contrast to 2D Monolayer Culture

In 2D monolayer culture, approximately 50% of the cells were in $S/G_2/M$ before chemotherapy, and after chemotherapy approximately 90% of the cells were blocked in $S/G_2/M$. In sphere culture, approximately 10% of the cells were in $S/G_1/M$ before chemotherapy, and after chemotherapy almost 100% of the cells were in G_0/G_1. In Gelfoam® histoculture, approximately 40% of the cells were in $S/G_0/M$ before chemotherapy, and during chemotherapy, almost 100% of the cancer cells were in G_0/G_1. After termination of chemotherapy approximately 20% of the cancer cells re-entered $S/G_2/M$, mostly at the surface on Gelfoam®, for both cisplatinum and paclitaxel. In the subcutaneous tumor, before chemotherapy approximately 30% of the cells were in $S/G_2/M$. During chemotherapy, almost 100% of the cancer cells were in G_0/G_1. After termination of chemotherapy, approximately 20% of the cancer cells re-entered $S/G_2/M$ for both cisplatinum and paclitaxel, mostly at the surface and very similar to Gelfoam® histocultures (Fig. 5). After the cessation of chemotherapy, the cancer cells at the surface of the tumor and Gelfoam® histocultures (*see* **Notes 4** and **5**) resumed cycling [3].

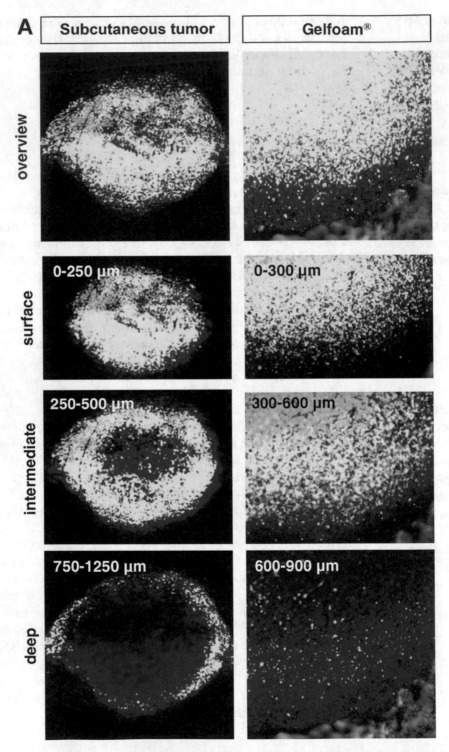

Fig. 3 Gelfoam®-histocultured tumor and subcutaneous FUCCI-expressing tumors have similar 3-dimensional spatial-temporal cell cycle phase distribution. (**a**) Representative images of FUCCI-expressing MKN45 cells in a tumor in the liver and tumor formed on Gelfoam® at the indicated depths. (**b**) Histograms show the cell-cycle distribution at the surface, intermediate area, and deep area of tumors in the liver and on Gelfoam®. Scale bars: 500 mm [3]

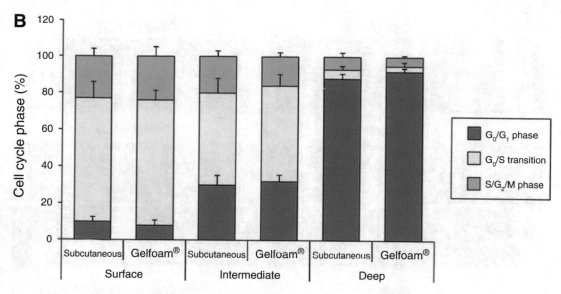

Fig. 3 (continued)

5   Notes

1. FUCCI imaging is unique as it can identify the phases of the cell cycle of each cell.

2. Cancer cells have a completely different cell cycle distribution in tumors and Gelfoam® compared to 2D monolayer culture.

3. The cell cycle phase of the cancer cell is the major determinant of whether the cell can respond to cytotoxic chemotherapy.

4. Drug response of cancer cells in Gelfoam® histoculture and tumors is similar due to a similar cell cycle distribution.

5. The present chapter demonstrates at the cellular level, enabled by FUCCI imaging, the similarity of behavior of cancer cells in Gelfoam® histoculture and in tumors which is very distinct from 2D culture.

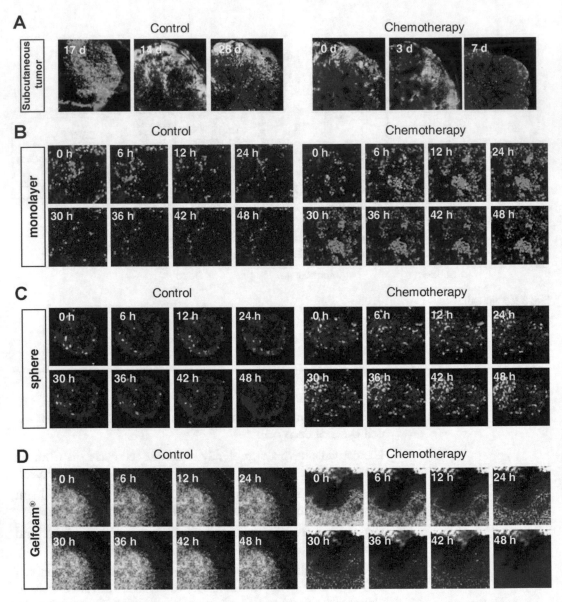

Fig. 4 FUCCI-expressing cancer cells on Gelfoam® have the same cell-cycle response to cytotoxic agents as subcutaneous tumors. Time-course imaging of FUCCI-expressing cancer cells: (**a**) subcutaneous, (**b**) in monolayer culture, (**c**) in tumor spheres on agar, and (**d**) on Gelfoam®, before and after chemotherapy. (**e**) Representative images of FUCCI-expressing cancer cells in monolayer culture, spheres on agar, on Gelfoam®, and subcutaneous tumors, before and after chemotherapy with cisplatinum or paclitaxel. In monolayer culture, chemotherapy blocked cancer cells in $S/G_2/M$ phase. Chemotherapy had little effect on quiescent tumor spheres. In contrast, tumors in Gelfoam® histoculture and subcutaneous tumors had a similar initial response to chemotherapy with cells becoming blocked in G_0/G_1. (F) Histograms of cell cycle phase distribution before and after chemotherapy of 2D monolayer, sphere, and Gelfoam® cultures and subcutaneous tumors [3]

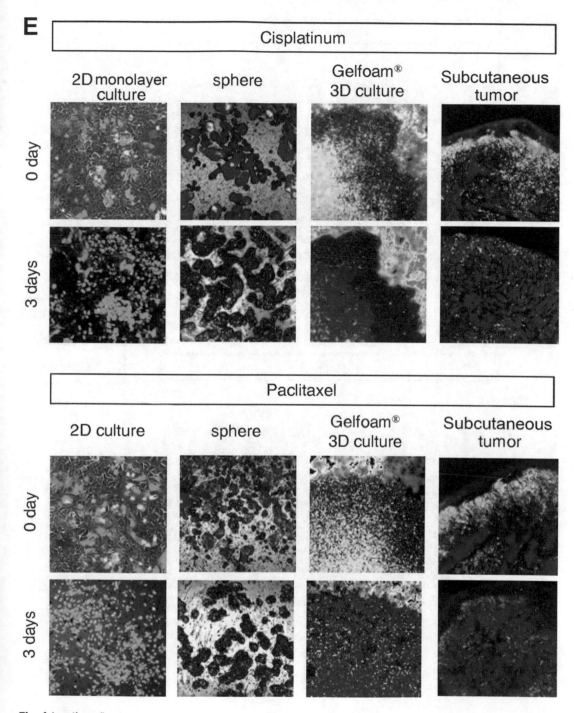

Fig. 4 (continued)

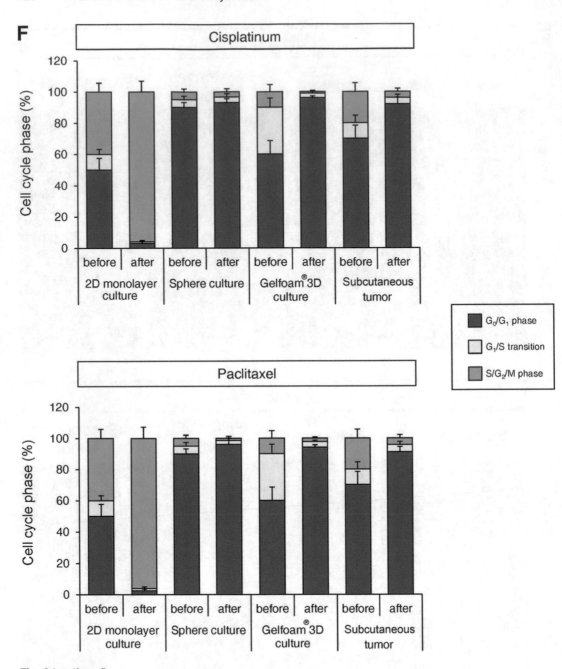

Fig. 4 (continued)

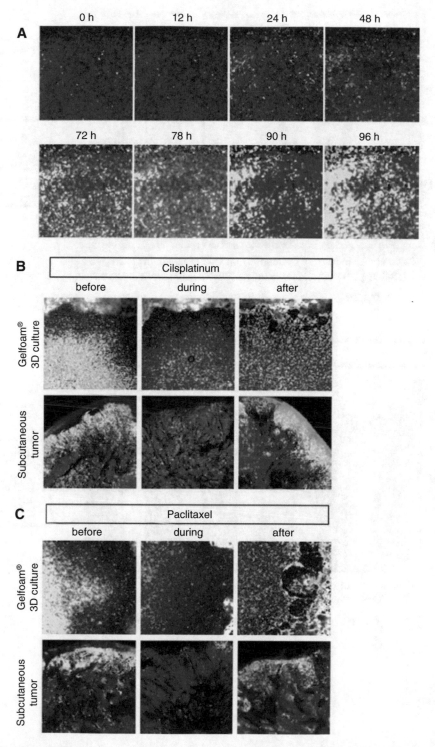

Fig. 5 FUCCI-expressing cancer cells on Gelfoam® have a similar recovery pattern from cisplatinum and paclitaxel as subcutaneous tumors. (**a**) Time-course imaging (96 h) of FUCCI-expressing MKN45 cells histocultured on Gelfoam® after cisplatinum chemotherapy. (**b**) Representative images of FUCCI-expressing cancer cells on Gelfoam® and subcutaneous tumors before and after recovery from chemotherapy with (**b, d**) cisplatinum or (**c, e**) paclitaxel. G_0/G_1 quiescent cancer cells on Gelfoam® and in subcutaneous tumors are resistant to chemotherapy and can restart proliferation after treatment is terminated. Scale bars: 500 mm [3]

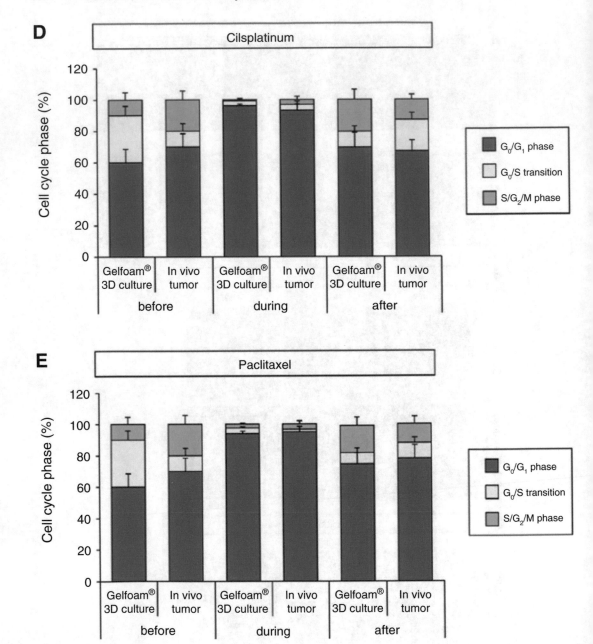

Fig. 5 (continued)

References

1. Sakaue-Sawano A, Kurokawa H, Morimura T, Hanyu A, Hama H, Osawa H, Kashiwagi S, Fukami K, Miyata T, Miyoshi H, Imamura T, Ogawa M, Masai H, Miyawaki A (2008) Visualizing spatiotemporal dynamics of multi-cellular cell cycle progression. Cell 132: 487–498

2. Yano S, Zhang Y, Miwa S, Tome Y, Hiroshima Y, Uehara F, Yamamoto M, Suetsugu A, Kishimoto H, Tazawa H, Zhao M, Bouvet M, Fujiwara T, Hoffman RM (2014) Spatial-temporal FUCCI imaging of each cell in a tumor demonstrates locational dependence of cell cycle dynamics and chemoresponsiveness. Cell Cycle 13:2110–2119

3. Yano S, Miwa S, Mii S, Hiroshima Y, Uehara F, Kishimoto H, Tazawa H, Zhao M, Bouvet M, Fujiwara T, Hoffman RM (2015) Cancer cells mimic *in vivo* spatial-temporal cell-cycle phase distribution and chemosensitivity in 3-dimensional Gelfoam® histoculture but not 2-dimensional culture as visualized with real-time FUCCI imaging. Cell Cycle 14:808–819

4. Yano S, Tazawa H, Hashimoto Y, Shirakawa Y, Kuroda S, Nishizaki M, Kishimoto H, Uno F, Nagasaka T, Urata Y, Kagawa S, Hoffman RM, Fujiwara T (2013) A genetically engineered oncolytic adenovirus decoys and lethally traps quiescent cancer stem-like cells into $S/G_2/M$ phases. Clin Cancer Res 19:6495–6505

5. Yano S, Miwa S, Mii S, Hiroshima Y, Uehara F, Yamamoto M, Kishimoto H, Tazawa H, Bouvet M, Fujiwara T, Hoffman RM (2014) Invading cancer cells are predominantly in G_0/G_1 resulting in chemoresistance demonstrated by real-time FUCCI imaging. Cell Cycle 13:953–960

6. Tome Y, Uehara F, Mii S, Yano S, Zhang L, Sugimoto N, Maehara H, Bouvet M, Tsuchiya H, Kanaya F, Hoffman RM (2014) 3-dimensional tissue is formed from cancer cells in vitro on Gelfoam®, but not on Matrigel™. J Cell Biochem 115:1362–1367

7. Freeman A, Hoffman RM (1986) In vivo-like growth of human tumors in vitro. Proc Natl Acad Sci U S A 83:2694–2698

8. Hoffman RM (2010) Histocultures and their use. In: Encyclopedia of life sciences. John Wiley and Sons Ltd, Chichester. https://doi.org/10.1002/9780470015902. a0002573

9. Hoffman RM (2013) Tissue culture. In: Brenner's Encyclopedia of Genetics, 2nd edn. Elsevier, Amsterdam, pp 73–76

10. Mii S, Duong J, Tome Y, Uchugonova A, Liu F, Amoh Y, Saito N, Katsuoka K, Hoffman RM (2013) The role of hair follicle nestin-expressing stem cells during whisker sensory-nerve growth in long-term 3-D culture. J Cell Biochem 114:1674–1684

Chapter 13

Methionine Dependency Determination of Human Patient Tumors in Gelfoam® Histoculture

Robert M. Hoffman

Abstract

The elevated requirement of methionine by cancer cells (methionine dependence) is a general metabolic abnormality in cancer. Methionine-dependent cancer cells are unable to proliferate and arrest in the late S/G_2 phase of the cell cycle when methionine is restricted in vitro or in vivo. Cell-cycle arrest in late S/G_2 was used as a biomarker of methionine dependence for patient tumors in Gelfoam® histoculture. Human cancer patient tumors, including tumors of the colon, breast, ovary, prostate, and a melanoma, were observed to be methionine dependent in Gelfoam® histoculture based on cell cycle analysis. This simple method can be used to screen patient tumors for methionine dependence and then subsequently apply appropriate chemotherapy for these patients to target this cancer-specific metabolic abnormality.

Key words Gelfoam® histoculture, Human tumors, Methionine dependence, Methionine restriction, Cell-cycle block, S/G_2 phase

1 Introduction

The elevated requirement of methionine by cancer cells is termed methionine dependence. There is much experimental support for this hypothesis that methionine dependence is a general metabolic defect in caner. Methionine dependence is due to excess use of methionine by cancer for aberrant transmethylation reactions that apparently divert methyl groups from DNA. The resulting global DNA hypomethylation is also a general phenomena in cancer. Global hypomethylation leads to an unstable genome and aneuploid karyotypes, another general phenomena in cancer. The excessive and aberrant use of methionine in cancer is observed clinically in [11C]methionine PET imaging, where high uptake of [11C]methionine results in a very strong and selective tumor signal compared with normal tissue background. [11C]methionine is superior to [18C] fluorodeoxyglucose (FDG)-PET for PET imaging, suggesting that methionine dependence is more tumor specific than glucose dependence [1, 2].

Robert M. Hoffman (ed.), *3D Sponge-Matrix Histoculture: Methods and Protocols*, Methods in Molecular Biology, vol. 1760, https://doi.org/10.1007/978-1-4939-7745-1_13, © Springer Science+Business Media, LLC, part of Springer Nature 2018

1.1 Enhanced Rates of Transmethylation in Cancer Cells

Cancer cells have enhanced overall rates of transmethylation compared with normal cells. The enhanced transmethylation rates may be the basis of the methionine dependence of cancer cells [3]. The elevated methionine use in cancer cells has been termed the "Hoffman effect" [4].

Methionine restriction of cancer cells resulted in their arrest in the S/G_2 phases of the cell cycle [5, 6].

1.2 Methionine-Independent Revertants

Rare cells from methionine-dependent cancer cell lines regained the normal ability to grow in methionine-restricted medium. These cell lines were termed methionine-independent revertants [7]. Methionine-independent revertants also had much lower basal transmethylation rates than parental methionine-dependent cell lines [8]. These results further demonstrated that methionine dependence is due to an increase in the rate of transmethylation reactions. Methionine-independent revertants cells concomitantly reverted for characteristics associated with cancer and became less malignant. Thus, the methionine-independent revertants become more normal-like indicating further a relationship between altered methionine metabolism and oncogenic transformation [9].

The hypothesis that methionine-dependence is due to excessive methionine use by cancer cells explains why the use of [^{11}C] methionine is so effective in positron-emission tomography (MET-PET) imaging, since the cancers use excessive methionine for their aberrant excess transmethylation and therefore take up excess [^{11}C] methionine, compared to normal tissue [10–14], as noted above. In a comparison of MET-PET and fluorodeoxyglucose (FDG)-PET, MET-PET was found to be superior for glioma [15] suggesting that cancer may have a greater abnormal requirement for methionine than glucose.

2 Materials

2.1 Preparation of Gelfoam® Sponges

1. Gelfoam® Absorbable Collagen Sponge USP, 12–7 mm format (Pfizer, NDC: 0009-0315-08) (Gelfoam®) as an example.

2. Gelfoam® trimming aid: Double-edged blade.

3. Sterile Petri dishes, 100 mm × 20 mm or wider.

4. Culture medium (CM): RPMI1640 with L-glutamine and phenol red, 100 μM modified Eagle's medium (MEM) nonessential amino acids (Thermo Fisher Scientific), 1 mM sodium pyruvate, 50 μg/mL gentamycin sulfate, 2.5 μg/mL amphotericin B (Thermo Fisher Scientific), 15% fetal bovine serum (FBS). We recommend testing different batches of FBS before purchasing a large stock.

5. Sterile metal forceps or tweezers.

6. Sterile metal scissors.

7. Sterile flat weighing metal spatula.

8. 6-Well culture plates.

9. Water-jacketed CO_2 incubator, set at 37 °C, 5% CO_2, ≥90% humidity.

10. Water bath.

11. Tissue transportation medium: Sterile phosphate buffer saline (PBS) or other solution with physiologic pH.

12. Sterile transportation container.

13. Tumor specimens from surgery.

14. Sterile Petri dish, 100 mm × 20 mm.

15. 70% ethanol solution.

16. Disinfectant solution for biological waste disposal.

17. 0.8% Collagenase.

18. Cytospin centrifuge.

19. 1 N hydrochloric acid

20. Schiff's reagent

21. Sulfurous acid

22. Phosphate-buffered saline

23. Cambridge Instruments Quantimet-520 system or equivalent

---

3   Methods

3.1   Histoculture

1. Obtain human patient tumors at the time of surgery.

2. Cut tissue into 1–3-mm^3 pieces and then place on Gelfoam® hydrated with culture medium.

3. Histoculture the tumor tissues in 6-well dishes and incubate in both methionine-repleted and methionine-depleted medium [16].

3.2   DNA Staining

1. Digest the histocultured tumor tissue along with supporting Gelfoam® with 0.8% collagenase for 2 h at different time points of histoculture.

2. Filter cell suspensions through a 200 μm mesh filter.

3. Wash the single-cell suspension with phosphate-buffered saline three times.

4. Spread the cells derived from the histocultures on slides with a cytospin centrifuge for DNA staining.

5. Incubate the slides containing the tumor cells in preheated 1 N hydrochloric acid at 60 °C for 15 min.

6. Rinse with distilled water six times to remove excess acid.

7. Stain with Schiff's reagent for 18 min at room temperature.

8. Treat with two successive changes of freshly prepared sulfurous acid rinse and wash in running tap water for 5 min.

9. Cover slip slides after dehydration for DNA analysis [16].

3.3 DNA Analysis

1. Use a Cambridge Instruments Quantimet-520 system or equivalent to determine the DNA content of cell nuclei by measuring the integrated optical density of each cell.

2. Measure the optical density of 500 cells on each slide. The integrated optical density is directly proportional to DNA content.

3. Use a control normal cell population as the standard [16].

4. Plot the number of cells as function of DNA content.

4 Results

4.1 Cell-Cycle Analysis of a Normal Cell Strain

The human foreskin fibroblast cell strain FS-3 was previously determined to be methionine independent by direct cell counting assays [17]. No difference was seen in the ratio of cells in G_1 to total cells as measured by DNA content after the cells were cultured in methionine-containing, homocysteine minus (MET+HCY-) medium or in methionine-restricted, homocysteine-containing (MET-HCY+) medium for 21 days. Normal human fibroblast strains do not change their cell cycle distribution value when grown in MET-HCY+ versus MET+HCY- media [5, 16] (*see* **Note 1**).

4.2 Cell-Cycle Analysis of Histocultured Patient Tumors

Tumors of the colon, ovary, prostate, and breast as well as melanoma were found to arrest in late S/G_2 under conditions of methionine restriction in Gelfoam® histoculture. Figure 1 shows the cell cycle distribution of all the methionine-dependent tumors in MET+HCY- compared to MET-HCY+ media demonstrating the late S/G_2 block of the Gelfoam®-histocultured tumors in methionine-depleted medium [16] (*see* **Notes 2 and 3**).

When the methionine-dependent patient melanoma was shifted from MET-HCY+ medium to MET+HCY- medium, a normal cell cycle distribution resumed, demonstrating the reversibility of the cell cycle block in the patient tumors [16] .Patient tumors demonstated to be methionine-dependent can be targeted by L-methionine α-deamino-γ-mercaptomethane lyase (recombinant methioniniase [rMETase] [EC 4.4.1.11]) [4, 18–27].

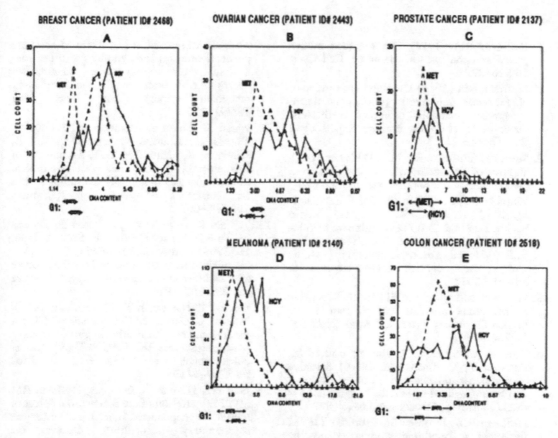

Fig. 1 Cell cycle distribution of human patient tumors shows them to be methionine dependent. Tumors were histocultured in MET⁺ HCY⁻ (MET) medium or in MET⁻ HCY⁺ (HCY) medium, stained with Schiff's reagent and measured for DNA content [16]. Note accumulation of cells with increased DNA content in MET⁻ HCY⁺ medium for all histocultured tumors shown, indicating an S/G₂-phase cell block thereby demonstrating methionine dependence [16]

---

5   Notes

1. We were able to grow patient tumors directly from surgery in Gelfoam® histoculture over a long term.

2. The results demonstrate that 3-dimensional Gelfoam® histo- culture and cell cycle analysis allow the determination of the methionine dependency of fresh human solid tumors [16].

3. These results suggest that the methionine-dependent tumors can be synchronized selectively with the late S/G₂ arrest induced by methionine restriction released by subsequent methionine repletion [5, 16].

References

1. Hoffman RM (2017) The wayward methyl group and the cascade to cancer. Cell Cycle 16:825–829

2. Hoffman RM (2015) Development of recombinant methioninase to target the general cancer-specific metabolic defect of methionine dependence: a 40-year odyssey. Expert Opin Biol Ther 15:21–31

3. Stern PH, Hoffman RM (1984) Elevated overall rates of transmethylation in cell lines from diverse human tumors. In Vitro 20:663–670

4. Murakami T, Li S, Han Q, Tan Y, Kiyuna T, Igarashi K, Kawaguchi K, Hwang HK, Miyaki K, Singh AS et al (2017) Recombinant methioninase effectively targets a Ewing's sarcoma in a patient-derived orthotopic xenograft (PDOX) nude-mouse model. Oncotarget 8: 35630–35638

5. Hoffman RM, Jacobsen SJ (1980) Reversible growth arrest in simian virus 40-transformed human fibroblasts. Proc Natl Acad Sci U S A 77:7306–7310

6. Yano S, Li S, Han Q, Tan Y, Bouvet M, Fujiwara T, Hoffman RM (2014) Selective methioninase-induced trap of cancer cells in S/G$_2$ phase visualized by FUCCI imaging confers chemosensitivity. Oncotarget 5:8729–8736

7. Hoffman RM, Jacobsen SJ, Erbe RW (1978) Reversion to methionine independence by malignant rat and SV40-transformed human fibroblasts. Biochem Biophys Res Commun 82:228–234

8. Judde JG, Ellis M, Frost P (1989) Biochemical analysis of the role of transmethylation in the methionine dependence of tumor cells. Cancer Res 49:4859–4865

9. Hoffman RM, Jacobsen SJ, Erbe RW (1979) Reversion to methionine independence in simian virus 40-transformed human and malignant rat fibroblasts is associated with altered ploidy and altered properties of transformation. Proc Natl Acad Sci U S A 76:1313–1317

10. Grosu AL, Weber WA, Riedel E, Jeremic B, Nieder C, Franz M, Gumprecht H, Jaeger R, Schwaiger M, Molls M (2005) L-(methyl-11C) methionine positron emission tomography for target delineation in resected high-grade gliomas before radiotherapy. Int J Radiat Oncol Biol Phys 63:64–74

11. Glaudemans AW, Enting RH, Heesters MA, Dierckx RA, van Rheenen RW, Walenkamp AM, Slart RH (2013) Value of 11C-methionine PET in imaging brain tumours and metastases. Eur J Nucl Med Mol Imaging 40:615–635

12. Tsuyuguchi N, Takami T, Sunada I, Iwai Y, Yamanaka K, Tanaka K, Nishikawa M, Ohata K, Torii K, Morino M et al (2004) Methionine positron emission tomography for differentiation of recurrent brain tumor and radiation necrosis after stereotactic radiosurgery–in malignant glioma. Ann Nucl Med 18:291–296

13. Nariai T, Tanaka Y, Wakimoto H, Aoyagi M, Tamaki M, Ishiwata K, Senda M, Ishii K, Hirakawa K, Ohno K (2005) Usefulness of L-[methyl-11C] methionine-positron emission tomography as a biological monitoring tool in the treatment of glioma. J Neurosurg 103:498–507

14. Tamura K, Yoshikawa K, Ishikawa H, Hasebe M, Tsuji H, Yanagi T, Suzuki K, Kubo A, Tsujii H (2009) Carbon-11-methionine PET imaging of choroidal melanoma and the time course after carbon ion beam radiotherapy. Anticancer Res 29:1507–1514

15. Singhal T, Narayanan TK, Jacobs MP, Bal C, Mantil JC (2012) 11C-methionine PET for grading and prognostication in gliomas: a comparison study with 18F-FDG PET and contrast enhancement on MRI. J Nucl Med 53:1709–1715

16. Guo HY, Herrera H, Groce A, Hoffman RM (1993) Expression of the biochemical defect of methionine dependence in fresh patient tumors in primary histoculture. Cancer Res 53:2479–2483

17. Mecham JO, Rowitch D, Wallace CD, Stern PH, Hoffman RM (1983) The metabolic defect of methionine dependence occurs frequently in human tumor cell lines. Biochem Biophys Res Commun 117:429–434

18. Tan Y, Xu M, Tan X, Tan X, Wang X, Saikawa Y, Nagahama T, Sun X, Lenz M, Hoffman RM (1997) Overexpression and large-scale production of recombinant L-methionine-alpha-deamino-gamma- mercaptomethane-lyase for novel anticancer therapy. Protein Expr Purif. 9:233–245

19. Takakura T, Ito T, Yagi S, Notsu Y, Itakura T, Nakamura T, Inagaki K, Esaki N, Hoffman RM, Takimoto A (2006) High-level expression and bulk crystallization of recombinant l-methionine g-lyase, an anticancer agent. Appl Microbiol Biotechnol 70:183–192

20. Takakura T, Takimoto A, Notsu Y, Yoshida H, Ito T, Nagatome H, Ohno M, Kobayashi Y, Yoshioka T, Inagaki K, Yagi S, Hoffman RM, Esaki N (2006) Physicochemical and pharmacokinetic characterization of highly potent recombinant l-methionine g-lyase conjugated with polyethylene glycol as an antitumor agent. Cancer Res 66:2807–2814

21. Takakura T, Misaki S, Yamashita M, Tamura T, Takakura T, Yoshioka T, Yagi S, Hoffman RM, Takimoto A, Esaki N, Inagaki K (2004) Assay method for antitumor l-methionine g-lyase: comprehensive kinetic analysis of the complex reaction with l-methionine. Anal Biochem 327:233–240

22. Kudou D, Misaki S, Yamashita M, Tamura T, Takakura T, Yoshioka T, Yagi S, Hoffman RM, Takimoto A, Esaki N, Inagaki K (2007) Structure of the antitumour enzyme l-methionine g-lyase from Pseudomonas putida at 1.8Å resolution. J Biochem 141:535–544

23. Kawaguchi K, Igarashi K, Li S, Han Q, Tan Y, Kiyuna T, Miyake K, Murakami T, Chmielowski B, Nelson SD, et al (2017) Combination treatment with recombinant methioninase enables temozolomide to arrest a BRAF V600E melanoma growth in a patient-derived orthotopic xenograft. Oncotarget 8:85516–85525

24. Kawaguchi K, Igarashi K, Li S, Han Q, Tan Y, Miyake K, Kiyuna T, Miyake M, Murakami T, Chmielowski S, et al (2018) Recombinant methioninase (rMETase) is an effective therapeutic for BRAF-V600E-negative as well as -positive melanoma in patient-derived orthotopic xenograft (PDOX) mouse models. Oncotarget 9:915–923

25. Igarashi K, Li S, Han Q, Tan Y, Kawaguchi K, Murakami T, Kiyuna T, Miyake K, Li Y, Nelson SD, et al (2018) Growth of a doxorubicin-resistant undifferentiated spindle-cell sarcoma PDOX is arrested by metabolic targeting with recombinant methioninase. J Cell Biochem, in press

26. Igarashi K, Kawaguchi K, Kiyuna T, Miyake K, Miyake M, Li S, Han Q, Tan Y, Zhao M, Li Y, et al (2018) Tumor-targeting Salmonella typhimurium A1-R combined with recombinant methioninase and cisplatinum eradicates an osteosarcoma cisplatinum-resistant lung metastasis in a patient-derived orthotopic xenograft (PDOX) mouse model: Decoy, trap and kill chemotherapy moves toward the clinic. Cell Cycle, in press

27. Kawaguchi K, Han Q, Li S, Tan Y, Igarashi K, Kiyuna T, Miyake K, Miyake M, Chmielowski B, Nelson SD, et al (2018) Targeting methionine with oral recombinant methioninase (o-rMETase) arrests a patient-derived orthotopic xenograft (PDOX) model of BRAF-V600E mutant melanoma: implications for clinical cancer therapy and prevention. Cell Cycle, in press

Chapter 14

Hair-Shaft Growth in Gelfoam® Histoculture of Skin and Isolated Hair Follicles

Robert M. Hoffman, Lingna Li, and Wenluo Cao

Abstract

Human scalp skin with abundant hair follicles in various stages of the hair growth cycle was histocultured for up to 40 days on Gelfoam® at the air/liquid interface. The anagen hair follicles within the histoculture scalp skin produced growing hair shafts. Hair follicles could continue their cycle in histoculture; for example, apparent spontaneous catagen induction was observed both histologically and by the actual regression of the hair follicle. In addition, vellus follicles were shown to be viable at day 40 after initiation of culture. Follicle keratinocytes continued to incorporate [³H]thymidine for up to several weeks after shaft elongation had ceased. Intensive hair growth was observed in the pieces of shaved mouse skin histocultured on Gelfoam®. Isolated human and mouse hair follicles also produced growing hair shafts. By day 63 in histoculture of mouse hair follicles, the number of hair follicle-associated pluripotent (HAP) stem cells increased significantly and the follicles were intact. Gelfoam® histoculture of skin demonstrated that the hair follicle cells are the most sensitive to doxorubicin which prevented hair growth, thereby mimicking chemotherapy-induced alopecia in Gelfoam® histoculture.

Key words Hair follicles, Chemotherapy, Hair growth, Human scalp, Mouse, Gelfoam® histoculture, Hair follicle-associate-pluripotent (HAP) stem cells

1 Introduction

In recent decades many attempts to establish in vitro systems of hair growth were undertaken [1–5]. Successful hair growth in isolated human hair follicles free-floating in culture medium was first reported by Philpott et al. [6–18]. Gelfoam® histoculture has been used to produce hair growth on human scalps, mouse skin, isolated human hair follicles, and isolated mouse follicles [7, 19–21].

Robert M. Hoffman (ed.), *3D Sponge-Matrix Histoculture: Methods and Protocols*, Methods in Molecular Biology, vol. 1760, https://doi.org/10.1007/978-1-4939-7745-1_14, © Springer Science+Business Media, LLC, part of Springer Nature 2018

2   Materials

2.1 Hair Follicles

1. Transgenic mice with nestin-driven GFP (ND-GFP) (4–8 weeks) (AntiCancer, Inc., San Diego, CA) as a source of whisker follicles [22, 23].

2. Human scalp skin from face lifts or scalp reduction or autopsy.

2.2 Imaging

1. Confocal laser scanning microscope (FV1000, Olympus Corp., Tokyo, Japan) for two- (X,Y) and three-dimensional (3D, X,Y,Z) high-resolution imaging with the 4×/0.10 Plan N, 10×/0.30 Plan-NEOFLUAR, 20×/0.50 UPlan FL N, and 20×/1.00w XLUMplan FL objectives [24].

2. Nikon or Olympus photomicroscope fitted with an epi-illumination polarization lighting system.

2.3 Preparation of Isolated Hair Follicles

2.3.1 Dissection or surgery equipment

1. Stereomicroscope.

2. Surgical knife.

3. Fine forceps.

4. Fine scissors.

5. Retractor for mice.

6. Syringe needles (10 ml, 5 ml, 1 ml syringe).

7. 31G insulin syringe.

2.3.2 Tissue culture supplies

1. Tissue and cell culture plates and flasks.

2. Collagen 1-coated 12-well and 48-well plates.

3. Collagen 1-coated 25 cm² flasks.

2.4 Histoculture

1. Sterile collagen sponges: Gelfoam® Absorbable Collagen Sponge USP, 12–7 mm format (Pfizer, NDC: 0009-0315-08).

2. Gelfoam® trimming aid: Double-edged blade.

3. Sterile Petri dishes, 100 mm × 20 mm or wider.

4. Culture medium (CM): RPMI1640 with L-glutamine and phenol red, 100 µM modified Eagle's medium (MEM) nonessential amino acids (Thermo Fisher Scientific), 1 mM sodium pyruvate, 50 µg/mL gentamycin sulfate, 2.5 µg/mL amphotericin B (Thermo Fisher Scientific), 15% fetal bovine serum (FBS). We recommend testing different batches of FBS before purchasing a large stock.

5. Sterile metal forceps or tweezers.

6. Sterile metal scissors.

7. Sterile flat weighing metal spatula.

8. 6-well culture plates.

9. Water-jacketed CO_2 incubator, set at 37 °C, 5% CO_2, ≥90% humidity.

10. Sterile Petri dish, 100 mm × 20 mm.

11. Ethanol solution (70%).

12. Disinfectant solution for biological waste disposal.

13. Kodak NTB-2 emulsion (Carestream Health, Rochester, NY).

14. Hematoxylin and eosin.

3 Methods

3.1 Preparation of Human Scalp Skin Tissue and Isolation of Human Hair Follicles

1. Pre-wash human scalp skin tissue taken from autopsy or facelift surgery in Earle's minimum essential medium with penicillin G, streptomycin, amphotericin B, tetracycline, amikacin, chloramphenicol, and gentamicin for 30 min.

2. Clean the epithelial surface of the scalp skin with 70% ethanol after the outgrowing hair shafts are shaved.

3. Cut intact scalp skin (2×2 cm^2) with a 2 mm Acu-punch or surgical blade along the direction of hair growth.

4. Carefully remove the additional subcutaneous fat with a scalpel without injury to the hair follicle bulbs [7].

5. Isolate intact individual human anagen hair follicles with a scalpel.

6. Dissect the isolated follicles at the infrainfundibular level without epidermis or sebaceous gland tissue, but with a narrow band of perifollicular connective tissue [7].

3.2 Histoculture of Human Scalp Tissue and Isolated Human Follicles

1. Explant small pieces of intact human scalp tissue (described above) with the epidermis up at the air/liquid interface and dermis-down on $1 \times 1 \times 1$ cm pieces on Gelfoam® that had been prehydrated for at least 4 h with culture medium (Eagle's minimum essential medium plus 10% fetal bovine serum and gentamicin at 50 µg/ml).

2. Maintain the cultures at 37 °C in a gassed incubator with 95% air/5% CO_2 [7].

3.3 [³H]thymidine Labeling of Proliferating Cells in the Skin and Hair Follicle

1. Incubate the Gelfoam® histocultures with [³H]thymidine (4 µCi/mL; 1 µCi = 37 kBq) for 3 days.

2. Wash with phosphate-buffered saline.

3. Fix with buffered 10% formalin and process for autoradiography as described [7, 19].

4. After exposure to the photographic emulsion and fixation, stain the slides with hematoxylin/eosin and analyze under epi-illumination polarization so that replicating cells can be

identified by the presence of silver grains over their nuclei, visualized as bright green in the epipolarization system [7, 19].

5. Dehydrate the cultures, embed in paraffin, and section by standard methods.

6. Deparaffinize the slides and prepare for autoradiography by coating with Kodak NTB-2 emulsion and expose for 5 days, after which they will develop [25–29].

7. Rinse and stain the slides with hematoxylin and eosin.

8. Analyze the slides and determine the percentage of cells undergoing DNA synthesis in treated vs. untreated tumor cultures using a Nikon or an Olympus photomicroscope fitted with epi-illumination polarization.

9. Identify replicating cells by the presence of silver grains, visualized as bright green in the epi-polarization system, over their nuclei due to exposure of the NTB-2 emulsion by radioactive DNA [28].

3.4 Histology and Hair Growth Measurement

1. Measure the length of hair shafts from photomicrographs [7].

2. Process tissue fixed in buffered 10% formalin for routine histology and stain with hematoxylin and eosin or Giemsa reagent according to standard procedures.

3.5 Isolation and Histocutlure of Mouse Vibrissa Follicles (Whiskers)

1. Conduct all animal studies are conducted in accordance with the principles and procedures outlined in the National Institute of Health Guide for the Care and Use of Animals under Assurance Number A3873-1.

2. Dissect the whisker pad to obtain isolated vibrissae follicles with forceps and fine needles, from mice anesthetized using a ketamine mixture (intramuscular injection of a 0.02 ml solution of 20 mg/kg ketamine, 15.2 mg/kg xylazine, and 0.48 mg/kg acepromazine maleate), using a binocular microscope (MZ6, Leica, Wetzlar, Germany).

3. Isolate vibrissae hair follicles with their capsules for Gelfoam® histoculture [22].

4. Place the vibrissae hair follicles in DMEM-F12 containing B-27 (2.5%), N2 (1%), and 1% penicillin and streptomycin on sterile Gelfoam® hydrated in the medium.

5. Incubate the Gelfoam®-histocultured follicles at 37 °C, 5% CO_2 100% humidity.

6. Change the medium every other day [22, 30].

3.6 Confocal Laser Scanning Microscopy

1. Use a confocal laser scanning microscope (FV1000) for two-(X,Y) and three-dimensional (3D, X,Y,Z) high-resolution imaging of the vibrissa follicles in Gelfoam® histoculture.

2. Obtain fluorescence images of green fluoresent protein (GFP) hair follicle-associated-pluripotent (HAP) stem cells in the hair follicle using the 4×/0.10 Plan N, 10×/0.30 Plan-NEOFLUAR, 20×/0.50 UPlan FL N, and 20×/1.00w XLUMplan FL objectives [24].

3.7 Toxicity Testing

1. To determine toxicity of chemotherapy or other agents, measure the inhibition of [³H]thymidine incorporation by histological autoradiography (*see* above) after doxorubicin treatment for 24 h.

2. Calculate the percentage of cell proliferation relative to control with a Nikon or an Olympus photomicroscope fitted with an epi-illumination polarization lighting system to. Image silver grains in autoradiographic preparations.

4 Results

When intact human scalp skin was cultured on Gelfoam® at the air/liquid interface, considerable elongation of pigmented and unpigmented hair shafts was observed over time (Fig. 1). The hair shafts grew an average of 0.86 ± 0.18 mm over 5 days and 1.10 ± 0.22 mm over 10 days. This correlated well with extensive proliferation of hair matrix keratinocytes during the first week of histoculture as seen by the cells covered by green grains in the autoradiographs (Fig. 2). The silver grains, which are formed in the photoemulsion over nuclei that have incorporated [³H]thymidine, reflect polarized light, which appears green. The follicle morphology was well preserved in histoculture, with the outer root sheath, matrix keratinocytes, dermal papilla, and hair shaft readily visible (Fig. 2) [7].

14–25% of follicles grown in intact skin showed hair shaft elongation within the first week of histoculture. In contrast, 73% of the isolated follicles explanted on Gelfoam® had hair shaft lengthening (*see* **Note 1**). The isolated follicles had been dissected at the infrainfundibular level; no epidermis or sebaceous gland tissue was present, only a narrow band of perifollicular connective tissue. The growth rate of the follicles in the intact skin histocultures was an average of 0.86 + 0.18 mm for 5 days, while the isolated follicles grew an average of 0.49 + 0.06 mm for 5 days [7].

Even after several weeks of histoculture of intact scalp skin, there was little follicle degeneration. Spontaneous regression (catagen) appeared to occur both in anagen hair follicles that either had been isolated and planted on Gelfoam® or were growing in intact human scalp skin histocultured on Gelfoam®. Follicle bulb shrinkage, increasing distance, as well as formation of a connective tissue strand between papilla and bulb, cessation of melanogenesis, and

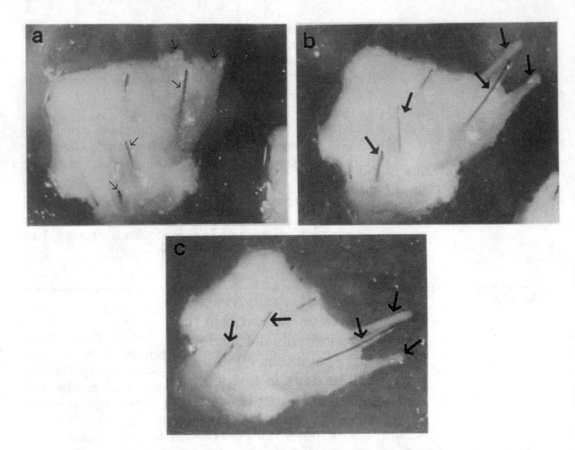

Fig. 1 Hair-shaft elongation in Gelfoam® histoculture of human scalp skin at day 0 (**a**) and after 5 days (**b**) and 10 days (**c**) of histoculture. Arrows point to elongating hair follicles. Note the increasing length of both dark and white hair (dissection microscopy, ×6) [7]

upward movement of the zone of pigmented keratinocytes in the follicle were observed indicating catagen (Figs. 3 and 4) [7].

In histocultured mouse skin, the length of the hairs increased with time of incubation (Fig. 5). The cells in the hair follicles not only were viable but also demonstrated an intensive DNA synthesis as determined by histological autoradiography of paraffin sections of specimens previously incubated with [³H]thymidine (Figs. 5 and 6) [20].

4.1 Hair-Shaft Elongation in Nestin-GFP-Expressing Follicles in Gelfoam® Histoculture

Isolated mouse whisker follicles from nestin-driven GFP transgenic mice with an intact capsule were then placed on Gelfoam® histoculture. Hair-shaft length in the follicles increased by 1.32 ± 0.27 mm by day 4 compared to day 1 and kept growing at day 7 (1.42 ± 0.24 mm) and day 9 (1.46 ± 0.24 mm) compared to day 1. By day 12, the hair-shaft length was 1.50 ± 0.22 mm ($p < 0.001$ compared to day 1) and remained constant until day 63. At day 63 of hair follicle histoculture, the ND-GFP-expressing

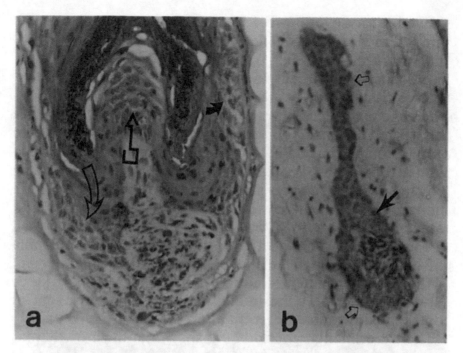

Fig. 2 Autoradiograph of [³H]thymidine incorporation into human scalp hair follicles after 6 days (anagen follicle) (**a**); vellus hair follicle after 40 days histoculture (**b**). Visualized by epipolarization, silver grains over radiolabeled nuclei reflect the polarized light as green. Note the remarkably preserved morphology of the hair follicles and the high proliferation frequency of the hair matrix cells and the outer-root-sheath follicle keratinocytes (**a**, ×700; **b**, ×350) [7]

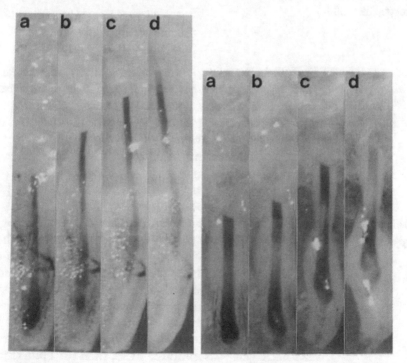

Fig. 3 Two representative examples of spontaneous regression of isolated human scalp hair follicles in Gelfoam® histoculture, suggestive of the onset of catagen. (**a**) Zero time. (**b**) Five days. (**c**) Ten days. (**d**) Fourteen days. Note the formation of club-shaped hair bulb and the movement of the hair shaft out of the follicle as a model of hair loss. (Dissection microscopy, ×15) [7]

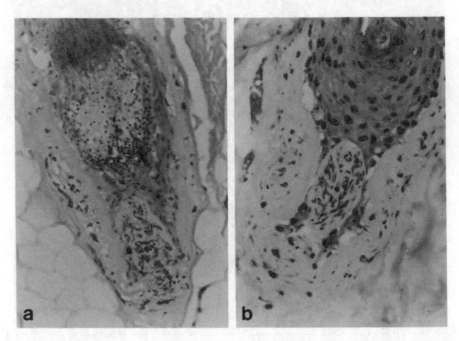

Fig. 4 Autoradiograph of [³H]thymidine incorporation into regressing human scalp hair follicles in Gelfoam®
histoculture. (**a**) Regressing follicle in intact scalp skin histocultured for 6 days. (**b**) Isolated follicle after 5 days
of histoculture. Note the increased distance between the dermal papilla and the follicle bulb, and the catagen-
characteristic formation of a tissue strand between the retreating, terminally differentiating cells of the bulb
and the dermal papilla (×600) [7]

hair follicle-associated pluripotent (HAP) stem cells (please *see*
Chap. 15 of the present volume) had a large increase in relative
fluorescence intensity and fluorescent area. The large increase in
ND-GFP expression of the HAP stem cells indicates their extensive
proliferation and activity, as well as the very-long-term viability of
the follicles in Gelfoam® histoculture. Stem cells increased over the
63-day histoculture period even though hair-shaft elongation
appeared to cease by day 20 (*see* **Notes 2** and **3**) [23].

**4.2   Hair Follicle
Chemotherapy Toxicity
in Gelfoam®
Histoculture**

[³H]thymidine incorporation and subsequent histological autoradiography in the hair follicle were used to indicate chemotherapy
toxicity in Gelfoam® histoculture. A 24-h exposure of histocultured skin to doxorubicin (DOX) inhibited [³H]thymidine incorporation of histocultured skin cells. The percentage of cell
proliferation relative to control was greatly decreased with increasing concentrations of DOX. The different cell types in skin had
different responses to DOX. [³H]thymidine incorporation in the
hair follicle cells was inhibited more than the epidermal or dermal
cells and, therefore, they are most sensitive to this cytotoxic agent,
thereby mimicking chemotherapy-induced alopecia histoculture
[31] (*see* **Note 4**).

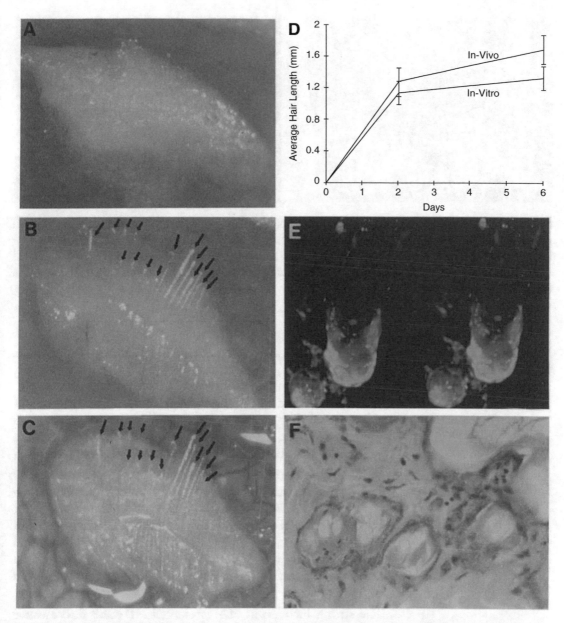

Fig. 5 (**a**) Shaved mouse skin histocultured on Gelfoam®. Note the absence of hair. Dissection microscopy. Magnification ×5.25. (**b**) Same specimen as (**a**) after histoculturing on Gelfoam for 2 days. Note the hair growing out from the skin. Dissection microscopy. Magnification ×5.25. (**c**) Same specimen as A after histoculturing on Gelfoam for 6 days. Note further hair growth. Dissection microscopy. Magnification ×5.25. (**d**) Comparison of the pattern of hair growth in vitro and in vivo versus time. Hair growth in vitro has high correlation with in vivo hair growth. (**e**) Hair follicles of mouse skin histocultured for 8 days and double-stained by BCECF-AM (green, living) and PI (red, dead). Note that most follicle cells are viable. Confocal scanning laser microscopy. Magnification < 350. (**f**) Autoradiography of paraffin section of shaved mouse skin histocultured for 6 days and labeled with [³H]thymidine for 4 days. Note the high frequency labeling of follicle cells (bright green grains). Polarized light and bright-field microscopy. Magnification ×500 [20]

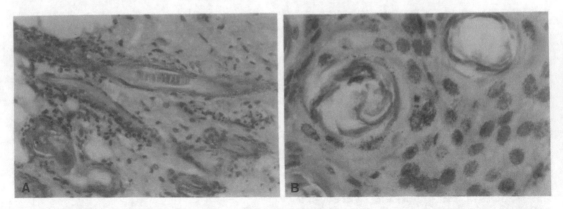

Fig. 6 (**a**) Mouse skin histocultured on Gelfoam® for 6 days in Eagle's minimum essential medium and labeled with [³H]thymidine for days 3–6. Tissue was fixed and processed for autoradiography and stained with hematoxylin and eosin. Autoradiograms were observed with bright-field and polarizing light. [³H]thymidine-labeled nuclei appear bright green. Note the high labeling index of hair follicle cells. (**b**) Same as (**a**). Note the labeled epidermal cells (**a:** ×1000; **b:** ×2000) [19]

5   Notes

1. Follicles dissected out of their surrounding skin showed a significantly higher percentage of hair-shaft elongation than those grown in their natural skin environment, although the follicles in the intact histocultured skin grew to a greater average length [7].

2. Isolated free-floating follicles were previously observed to produce elongating hair shafts but apparently were not viable for very long periods of time [6, 32]. Gelfoam® supports both hair shaft and nerve growth [22] of isolated whisker follicles and maintained viability of the follicles for at least 63 days, much longer than free-floating follicles, which can enable long-term experimentation.

3. The strong increase in HAP stem cell ND-GFP fluorescence indicates increased activity and proliferation of the stem cells. The greatest increase in HAP stem cell fluorescence is during the period of hair-shaft elongation [23].

4. In experiments using inhibition of [³H]thymidine as an end point, we have shown that DOX preferentially inhibits follicle cells, which mirrors the in vivo effect of DOX. Inhibition of [³H]thymidine incorporation may provide an intermediate end point to measure the toxicity of substances at high levels of sensitivity.

References

1. Strangeways DH (1931) The growth of hair in vitro. Arch Exp Zellforsch 11:344

2. Murray MR (1933) Development of the hair follicle and hair in vitro. Anat Rec 57(Supplement):74

3. Hardy MH (1951) The development of pelage hairs and vibrissae from skin in tissue culture. Ann N Y Acad Sci 53:546–561

4. Frater R, Whitmore PG (1973) The *in-vitro* growth of post-embryonic hair. J Invest Dermatol 61:72–81

5. Philpott M, Green M, Kealey T (1989) Studies on the biochemistry and morphology of freshly isolated and maintained rat hair follicles. J Cell Sci 93:409–418

6. Philpott M, Green M, Kealey T (1990) Human hair growth in vitro. J Cell Sci 97:463–471

7. Li L, Paus R, Margolis LB, Hoffman RM (1992) Hair shaft elongation, follicle growth, and spontaneous regression in long-term, gelatin sponge-supported histoculture of human scalp skin. Proc Natl Acad Sci U S A 89:8764–8768

8. Kido T, Horigome T, Uda M, Adachi N, Hirai Y (2017) Generation of iPS-derived model cells for analyses of hair shaft differentiation. BioTechniques 63:131–134

9. Ojeh N, Akgül B, Tomic-Canic M, Philpott M, Navsaria H (2017) In vitro skin models to study epithelial regeneration from the hair follicle. PLoS One 12:e0174389

10. Su YS, Fan ZX, Xiao SE, Lin BJ, Miao Y, Hu ZQ, Liu H (2017) Icariin promotes mouse hair follicle growth by increasing insulin-like growth factor 1 expression in dermal papillary cells. Clin Exp Dermatol 42:287–294

11. Chacón-Martínez CA, Klose M, Niemann C, Glauche I, Wickström SA (2017) Hair follicle stem cell cultures reveal self-organizing plasticity of stem cells and their progeny. EMBO J 36:151–164

12. Zhou L, Yang K, Xu M, Andl T, Millar SE, Boyce S, Zhang Y (2016) Activating β-catenin signaling in CD133-positive dermal papilla cells increases hair inductivity. FEBS J 283:2823–2835

13. Hwang I, Choi KA, Park HS, Jeong H, Kim JO, Seol KC, Kwon HJ, Park IH, Hong S (2016) Neural stem cells restore hair growth through activation of the hair follicle niche. Cell Transplant 25:1439–1451

14. Paus R, Burgoa I, Platt CI, Griffiths T, Poblet E, Izeta A (2016) Biology of the eyelash hair follicle: an enigma in plain sight. Br J Dermatol 174:741–752

15. Langan EA, Philpott MP, Kloepper JE, Paus R (2015) Human hair follicle organ culture: theory, application and perspectives. Exp Dermatol 24:903–911

16. Sohn KM, Jeong KH, Kim JE, Park YM, Kang H (2015) Hair growth-promotion effects of different alternating current parameter settings are mediated by the activation of Wnt/β-catenin and MAPK pathway. Exp Dermatol 24:958–963

17. Kiso M, Hamazaki TS, Itoh M, Kikuchi S, Nakagawa H, Okochi H (2015) Synergistic effect of PDGF and FGF2 for cell proliferation and hair inductive activity in murine vibrissal dermal papilla in vitro. J Dermatol Sci 79:110–118

18. Yang XY, Jin K, Ma R, Yang JM, Luo WW, Han Z, Cong N, Ren DD, Chi FL (2015) Role of the planar cell polarity pathway in regulating ectopic hair cell-like cells induced by Math1 and testosterone treatment. Brain Res 1615:22–30

19. Li L, Margolis LB, Hoffman RM (1991) Skin toxicity determined *in vitro* by three-dimensional, native-state histoculture. Proc Natl Acad Sci U S A 88:1908–1912

20. Li L, Paus R, Margolis LB, Hoffman RM (1992) Hair growth *in vitro* from histocultured skin. In Vitro Cell Dev Biol 28A:479–481

21. Li L, Paus R, Slominski A, Hoffman RM (1992) Skin histoculture assay for studying the hair cycle. In Vitro Cell Dev Biol 28A:695–698

22. Mii S, Duong J, Tome Y, Uchugonova A, Liu F, Amoh Y, Hoffman RM (2013) The role of hair follicle nestin-expressing stem cells during whisker sensory-nerve growth in long-term 3D culture. J Cell Biochem 114:1674–1684

23. Cao W, Li L, Mii S, Amoh Y, Liu F, Hoffman RM (2015) Extensive hair shaft elongation by isolated mouse whisker follicles in very long-term Gelfoam® histoculture. PLoS One 10:e0138005

24. Uchugonova A, Zhao M, Weinigel M, Zhang Y, Bouvet M, Hoffman RM, Koenig K (2013) Multiphoton tomography visualizes collagen fibers in the tumor microenvironment that maintain cancer-cell anchorage and shape. J Cell Biochem 114:99–102

25. Jahoda CAB, Home KA, Oliver RF (1984) Induction of hair growth by implantation of cultured dermal papilla cells. Nature 311:560–562

26. Rogers G, Martinet N, Steinert P, Wynn P, Roop D, Kilkenny A, Morgan D, Yuspa SH (1987) Cultivation of murine hair follicles as organoids in a collagen matrix. J Invest Dermatol 89:369–379

27. Paus R, Stenn KS, Link RE (1989) The induction of anagen hair growth in telogen mouse skin by cyclosporine A administration. Lab Investig 60:365–369

28. Paus R, Stenn KS, Link RE (1990) Telogen skin contains an inhibitor of hair growth. Br J Dermatol 122:777–784

29. Paus R, Stenn KS, Elgjo K (1991) The epidermal pentapeptide pyroGlu-Glu-Asp-Ser-GlyOH inhibits murine hair growth in vivo and in vitro. Dermatologica 183:173–178

30. Duong J, Mii S, Uchugonova A, Liu F, Moossa AR, Hoffman RM (2012) Real-time confocal imaging of trafficking of nestin-expressing multipotent stem cells in mouse whiskers in long-term 3-D histoculture. In Vitro Cell Dev Biol Anim 48:301–305

31. Goldberg MT, Tackaberry LE, Hardy MH, Noseworthy JH (1990) Nuclear aberrations in hair follicle cells of patients receiving cyclophosphamide. A possible in vivo assay for human exposure to genotoxic agents. Arch Toxicol 64:116–121

32. Jo SJ, Choi S-J, Yoon S-Y, Lee JY, Park W-S, Park P-J et al (2013) Valproic acid promotes human hair growth in in vitro culture model. J Dermatol Sci 72:16–24

Chapter 15

Hair Follicle-Associated Pluripotent (HAP) Stem Cells in Gelfoam® Histoculture for Use in Spinal Cord Repair

Fang Liu and Robert M. Hoffman

Abstract

The stem cell marker, nestin, is expressed in the hair follicle, both in cells in the bulge area (BA) and the dermal papilla (DP). Nestin-expressing hair follicle-associated-pluripotent (HAP) stem cells of both the BA and DP have been previously shown to be able to form neurons, heart muscle cells, and other non-follicle cell types. The ability of the nestin-expressing HAP stem cells from the BA and DP to repair spinal cord injury was compared. Nestin-expressing HAP stem cells from both the BA and DP grew very well on Gelfoam®. The HAP stem cells attached to the Gelfoam® within 1 h. They grew along the grids of the Gelfoam® during the first 2 or 3 days. Later they spread into the Gelfoam®. After transplantation of Gelfoam® cultures of nestin-expressing BA or DP HAP stem cells into the injured spinal cord (including the Gelfoam®) nestin-expressing BA and DP cells were observed to be viable over 100 days post-surgery. Hematoxylin and eosin (H&E) staining showed connections between the transplanted cells and the host spine tissue. Immunohistochemistry showed many Tuj1-, Isl 1/2, and EN1-positive cells and nerve fibers in the transplanted area of the spinal cord after BA Gelfoam® or DP Gelfoam® cultures were transplanted to the spine. The spinal cord of mice was injured to effect hind-limb paralysis. Twenty-eight days after transplantation with BA or DP HAP stem cells on Gelfoam® to the injured area of the spine, the mice recovered normal locomotion.

Key words Hair follicle, Bulge area, Dermal papilla, Nestin, HAP stem cells, Gelfoam®, Spinal cord injury, Repair

1 Introduction

The hair follicle has recently been shown to be a source of nestin-expressing pluripotent stem cells that can form neurons, heart muscle cells, and other non-follicular cell types [1–8]. We have termed these cells hair follicle-associated pluripotent (HAP) stem cells. HAP stem cells are located in the bulge area (BA) surrounding the hair shaft and are interconnected by short dendrites. HAP stem cells were also found in the dermal papilla (DP) [9].

Robert M. Hoffman (ed.), *3D Sponge-Matrix Histoculture: Methods and Protocols*, Methods in Molecular Biology, vol. 1760, https://doi.org/10.1007/978-1-4939-7745-1_15, © Springer Science+Business Media, LLC, part of Springer Nature 2018

Bulge explants from adult mouse whisker follicles containing HAP stem cells gave rise to cells which formed neurons, heart muscle cells, smooth muscle cells, Schwann cells, and melanocytes [2–4, 8], thereby confirming the observation of Li et al. [1], who discovered these cells.

When nestin-expressing HAP stem cells from the mouse vibrissa BA were implanted into the gap region of the severed sciatic nerve of mice, they greatly enhanced the rate of nerve regeneration and the restoration of nerve function. The transplanted follicle BA cells transdifferentiated largely into Schwann cells, which are known to support neuron regrowth. The transplanted mice recovered the ability to walk normally [9].

Nestin-expressing mouse vibrissa BA HAP stem cells were transplanted to the injury site of C57B16 mice in which the thoracic spinal cord was severed. Most of the transplanted cells also differentiated into Schwann cells that apparently facilitated repair of the severed spinal cord. The rejoined spinal cord re-established extensive hind-limb locomotor performance [10].

Yu et al. [11, 12] isolated a population of human nestin-expressing hair follicle BA cells. These cells also expressed neural crest and neuron stem cell markers as well as the embryonic stem cell transcription factors Nanog and Oct4. The human BA cells proliferated as spheres, were capable of self-renewal, and differentiated into multiple lineages including myogenic, melanocytic, and neuronal cell lineages after in vitro clonal single-cell culture. In addition, the human nestin-expressing BA cells differentiated into adipocyte, chondrocyte, and osteocyte lineages. The human BA cells shared a similar gene expression pattern with murine skin immature neural crest cells [11, 12].

Human nestin-expressing HAP stem cells from the BA were transplanted in the severed sciatic nerve of the mouse where they differentiated into glial fibrillary acidic protein (GFAP)-positive Schwann cells and promoted the recovery of preexisting axons, leading to nerve regeneration. The regenerated nerve recovered function and, upon electrical stimulation, contracted the gastrocnemius muscle [13].

This chapter reviews results demonstrating that Gelfoam® histocultures of nestin-expressing HAP stem cells, fromthe BA and DP, that can affect spinal cord injury repair [14].

2 Materials

2.1 Whisker and Spinal Surgery Components

1. Animals: Nestin-driven GFP (ND-GFP) transgenic mice for BA culture, RFP transgenic mice for DP culture, non-transgenic nude mice for spinal surgery.

2. HAP stem cell sources: Whisker-follicle BA or DP.

3. Dissection or surgery equipment: Stereomicroscope, surgical knife, fine forceps, fine scissors, retractor for mice, syringe needles (10 ml, 5 ml, 1 ml syringe), 31G insulin syringe.

4. Tissue and cell culture plate or flask: Tissue culture dishes (60 mm), collagen I-coated 12-well and 48-well plates, collagen I-coated 25 cm² flasks.

5. Culture medium: Dulbecco's modified Eagle's medium (DMEM)/F12 (GIBCO, Grand Island, NY); 10% fetal bovine serum (FBS) (GIBCO); 1% N2 (GIBCO); 2% B27 (GIBCO); 200 mM L-glutamine (GIBCO); 0.025% ITS +3 (Sigma Aldrich, St. Louis, MD); 20 ng/ml epithelial growth factor; 20 ng/ml basic fibroblast growth factor (Invitrogen, Carlsbad, CA). Subculture medium is similar to the culture medium but serum free.

6. Phosphate-buffered saline (PBS)

7. Trypsin.

2.2 Gelfoam® Preparations

1. Sterile collagen sponges: Gelfoam® Absorbable Collagen Sponge USP, 12–7 mm format (Pfizer, NDC: 0009-0315-08).

2. Gelfoam® trimming aid: double-edged blade.

2.3 Imaging

1. FV1000 confocal microscope (Olympus, Tokyo, Japan) or equivalent.

2. OV100 Small Animal Imaging System (Olympus, Tokyo, Japan) or equivalent.

3. IV100 Laser Scanning Microscope (Olympus, Tokyo, Japan) or equivalent.

2.4 Immunofluorescence Reagents

2.4.1 Primary Antibodies Used are:

1. Nestin (rabbit, 1:100) (Sigma).
2. Versican (mouse, 1:100) (Millipore).
3. Tuj1 (rabbit, 1:30) (Sigma).
4. GFAP (rabbit, 1:100) (Sigma).
5. Isl 1/2 (rabbit, 1:100) (Santa Cruz).
6. EN-1 (rabbit, 1:100) (Santa Cruz).

2.4.2 Secondary Antibodies Used are:

1. Goat anti-rabbit IgG FITC (Santa Cruz).
2. Donkey anti-rabbit IgG TRITC (Santa Cruz).
3. Donkey anti-mouse IgG TRITC (Santa Cruz) (1:100)

2.4.3 Other reagents used are:

1. Triton X-100.
2. 5% normal goat serum.
3. DAPI, 1:1000.
4. Fluoromount (Sigma).
5. Paraformaldehyde.

3 Methods

3.1 Culture of HAP Stem Cells from the BA and DP of Mouse Vibrissa Follicles

1. Mouse whisker pad dissection: Anesthetize ND-GFP or RFP transgenic mice with tribromoethanol. Spray 75% ethanol on the whisker area (*see* **Note 1**) and cut off the whisker pad with ophthalmic scissors (*see* **Note 2**). Suture the wounds and keep the mice in an animal warming chamber until they recover from anesthesia.

2. Hair follicle isolation: Place the whisker pads in 75% ethanol for further sterilization for 3 min and wash in 0.01 M PBS three times, 5 min each. Expose the inner surface of the whisker pad and cut it into several strips parallel to the hair follicle. Separate individual hair follicles within the hair follicle strips with fine scissors or 5 ml syringe needles under a stereomicroscope. Choose anagen follicles and wash in PBS for further dissection (*see* **Note 3**).

3. Dissection of BA and DP from the hair follicle: Cut the anagen follicle into an upper part and a lower part (*see* **Note 4**). Put BA and DP in different dishes with PBS. The upper part of the hair follicle contains the BA and hair shaft and the lower part contains the DP. Tear off the outer connective capsule of the two parts carefully with 1 ml syringe needles. Expose the BA and the DP (*see* **Note 5**). Wash the BA and DP in PBS three times.

4. Culture of HAP stem cells from the BA and DP: Put the BA or DP into collagen 1-coated 12-well plates. Arrange the BAs or DPs evenly on the plate, 5–8 in each well (*see* **Note 6**). Place the plates with BAs or DPs without culture medium in an incubator (37 °C, 95% CO_2, 5% O_2) for 40 min. Then add 200 μl primary culture medium as slow as possible along the wall of each well.

5. Subculture of HAP stem cells: Remove the BA or DP carefully with a 1 ml syringe needle after 7 days of primary culture. Rinse the plate with PBS three times. Incubate the BA- or DP-derived HAP stem cells with 0.25% trypsin. Use trypsin inhibitor to terminate the digestion after 3–5 min. Add 800 μl culture medium (serum free) to each well. Pipette the medium gently 20 times with a 1 ml pipette tip. Transfer the medium from plate to a 10 ml centrifuge tube and centrifuge at 1000 rpm for 5 min. Discard the supernatant carefully. Add 5 ml serum-free culture medium to the tube and pipette gently to obtain a cell suspension. Put the cell suspension in a collagen 1-coated 25 cm^2 flask for subculture. Renew half of the medium every 2–3 days.

3.2 Three-Dimensional Culture of HAP Stem Cells on Gelfoam®

1. Harvesting of the subcultured HAP stem cells: Rinse the HAP stem cells in the flask with PBS and then treat the cells with 0.25% trypsin. Harvest the cell suspension with the same protocols as for subculture of HAP stem cells described above. Count the cells with a hemocytometer and adjust the cell density to 1.0×10^7 cells/ml. Transfer the cell suspension to a 500 µl Eppendorf tube and keep it on ice until use.

2. Construction of the Gelfoam®-HAP stem cell complex: Trim the Gelfoam® sponge into 3 mm³ cubes with a double-edged blade. Inject the HAP stem cell suspension into the Gelfoam® at 1 mm intervals with a 31G insulin syringe under a stereomicroscope. A total of 20 µl cell suspension is injected in one Gelfoam® cube (*see* **Note 7**).

3. Culture of the Gelfoam®-HAP stem cell complex: Place the Gelfoam®-HAP stem cell complex in a 48-well plate with no medium and put it in an incubator (37 °C, 95% CO_2, 5% O_2) for 1 h. Then add 1 ml subculture medium to each well gently from the wall. Renew half of the medium every 2–3 days. Culture the complex for 10 days until use.

3.3 Surgical Procedures for Spinal Cord Injury

1. Exposure of the spine and spinal cord: Anesthetize the nude mice with tribromoethanol. Place the mouse upright. Spray 75% ethanol on the back skin for sterilization. Cut the skin overlaying the thoracic vertebral column with a surgical knife. Separate the muscles along the spine with fine scissors. Expose the thoracic vertebra with a retractor. Separate the joints of the vertebra. Perform the laminectomy at T8 level (*see* **Note 8**) and expose the corresponding spinal cord segment. Rinse the separated vertebral lamina in PBS.

2. Semi-transection injury of the spinal cord: Cut the right side of the spinal cord transversely from medial to lateral with a 31G insulin syringe needle (*see* **Note 9**). The surface size of the lesion is 500 µm × 300 µm. The depth of the lesion is from the dorsal to the abdominal surface of the spinal cord. Use wet Gelfoam® soaked in PBS to clean the lesion site three times.

3.4 Transplantation of the Gelfoam®-HAP Stem Cell Complex

1. Wash the Gelfoam®-HAP stem cell complex with PBS three times. Cut it into three parts evenly with a double-edged blade (*see* **Note 10**). Place one part into the lesion site on the spinal cord. Cover the vertebral lamina back on the lesion site. Suture the wound and keep the mice in a warming chamber until they recover from anesthesia.

3.5 Immunocyto-chemistry and immunohisto-chemistry

1. Fix cultured BA and DP cells or frozen sections from the spinal cord in pre-cooled 4% paraformaldehyde at room temperature (RT) for 10 minutes.

2. Wash with PBS three times.

3. Treat slides with 0.3% Triton X-100 and incubate at RT for 30 minutes.

4. Apply 5% normal goat serum at RT for 30 minutes.

5. Apply primary antibodies at 4°C for 48 hours.

6. Apply secondary antibodies at RT in the dark for 2.5 hours.

7. Dilute DAPI, 1:1000, in PBS and stain at RT for 5 minutes.

8. Mount slides with Fluoromount and observe under fluorescence microscopy.

4 Results

4.1 Localization of Nestin-Expressing HAP Stem Cells in the Vibrissa BA and DP

In order to visualize the localization of nestin-GFP-expressing HAP stem cells in the mouse hair follicle, the anagen vibrissa follicle from ND-GFP mice (8 weeks) was dissected. The follicle was covered by a rigid capsule. It had two blood-filling areas (Fig. 1A1). One was at the BA, and the other was at the bulb area, which contained the DP. The BA and DP could be clearly visualized after the capsule was removed (Fig. 1A2–A7). Very bright GFP fluorescence was found at the BA and DP using the OV100 Small Animal Imaging System (Fig. 1A3, A5, A7), indicating that nestin-expressing HAP stem cells were in these two regions [14].

4.2 Nestin Expression in the BA and DP Varies with Follicle Stage

To understand whether the expression of nestin in the BA and DP was related to animal age or follicle stage, vibrissa follicles were dissected from 2-week- and 9-month-old ND-GFP mice. The BA in all hair follicle stages expressed nestin very strongly in animals of both ages. In middle anagen follicles (Fig. 2B1–B4, C1–C4, Fig. 3B1, B2), the area between the BA and DP extended and nestin-expressing blood vessels were present. Some blood vessels were directly associated with the DP (Fig. 2C4). The DP had the most extensive nestin expression during middle anagen (Fig. 2B1–B4, C1–C4). In late anagen follicles, the DP lost nestin expression (Fig. 2D1–D4). Similar patterns were seen with 9-month-old mice (data not shown) [14].

4.3 Nestin-Expressing HAP Stem Cells from BA and DP Have Similar Morphology

Confocal microscopy was used to visualize individual nestin-expressing HAP stem cells in the DP and BA. Nestin-expressing oval-shaped HAP stem cells with long dendrite-like structures were observed in both the BA and DP within isolated vibrissa follicles (Fig. 3a, b). Oval-shaped HAP stem cells with long dendrite-like structures were observed from the BA and DP when the cells were cultured on Gelfoam® (Fig. 3c, d). The HAP stem cells from both the BA and DP were essentially identical in morphology [14].

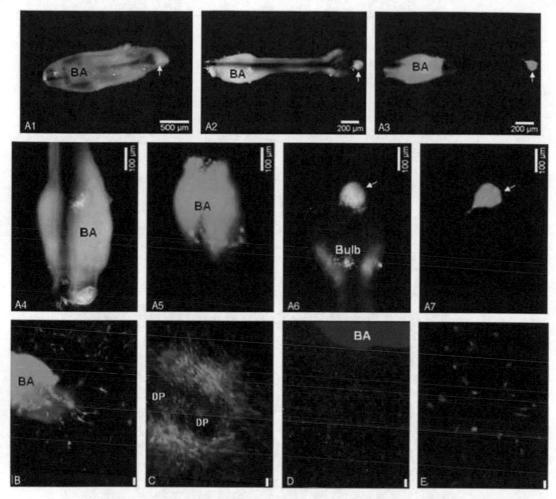

Fig. 1 Nestin-expressing HAP stem cells in the bulge area (BA) and dermal papilla (DP) of mouse vibrissal follicle. (**A1–A3**) Whole view of an ND-GFP follicle. The follicle capsule was intact in **A1**. In **A2** and **A3**, the capsule was removed. The BA and DP had strong nestin-GFP fluorescence. **A4–A7** are the higher magnification of **A2** and **A3**. The DP is shown by arrows. (**B**) Primary culture of BA ND-GFP HAP stem cells. (**C**) Primary culture of DP ND-GFP HAP stem cells. Two ND-GFP DP are shown. (**D**) Primary culture of RFP BA cells. (**E**) Nestin immunofluorescence staining of RFP BA cells. The BA cells were nestin positive. (**F**) Bulge area. (**G**) Dermal papilla. Panels **A1–A7** were observed with the Olympus OV100 Small-Animal Imaging System. Panels **B–E** were observed under fluorescence microscopy. Scale bar in **A1**: 500 μm. Scale bars in **A2** and **A3**: 200 μm. Scale bars in **A4–A7**: 100 μm. Scale bars in **B, D, E**: 25 μm. Scale bar in **C**: 50 μm [14]

4.4 Nestin-Expressing HAP Stem Cells from Both BA and DP Can Form Spheres

Nestin-expressing HAP stem cells from the BA and DP were cultured in serum-free medium supplemented with β-FGF and EGF (Fig. 1B, C). After primary culture for 4–7 days, nestin-expressing HAP stem cells grew out from the BA and DP. BA cells derived from whiskers from red fluorescent protein (RFP)-expressing mice were cultured using the same medium as that for BA ND-GFP HAP stem cells and DP cells (Fig. 1D). Immunocytochemistry demonstrated that the RFP-BA cells also expressed nestin (Fig. 1E). HAP

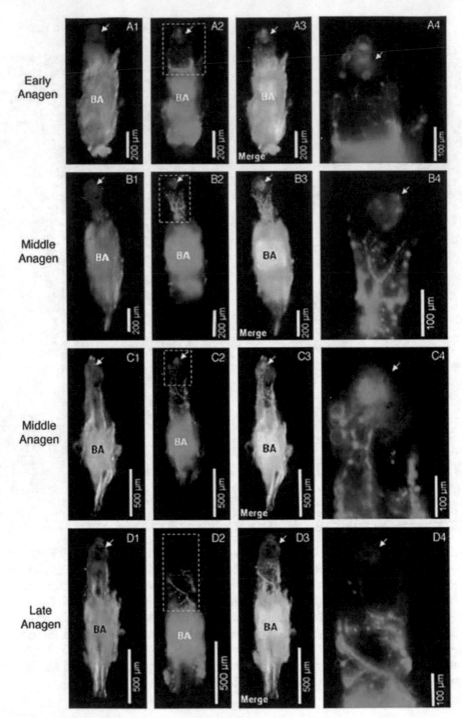

Fig. 2 Vibrissa follicles from 2-week-old nestin-GFP mouse. The capsule of the follicles was removed. The BA in all the follicles expressed nestin-GFP very strongly. Panels (**A1–A4**) show follicles in early anagen. The area between BA and DP was small and the blood vessels in this area were not yet formed. Panels (**B1–B4**) and (**C1–C4**) show follicles in middle anagen. The area between the BA and DP extended and blood vessels were formed. Some blood vessels connected to the DP directly (**C4**). The DP had strong nestin-GFP expression. Panels (**D1–D4**) show a follicle in late anagen. The DP lost nestin-GFP fluorescence. Imaging was with the Olympus OV100. Arrows point to the DP. Panels (**A4, B4, C4, D4**) are the higher magnification of dotted squares in (**A2, B2, C2, and D2**), respectively. Scale bars in (**A1–A3** and **B1–B3**): 200. Scale bars in (**C1–C3** and **D1–D3**): 500 μm. Scale bars in (**A4, B4, C4, and D4**): 100 μm [14]

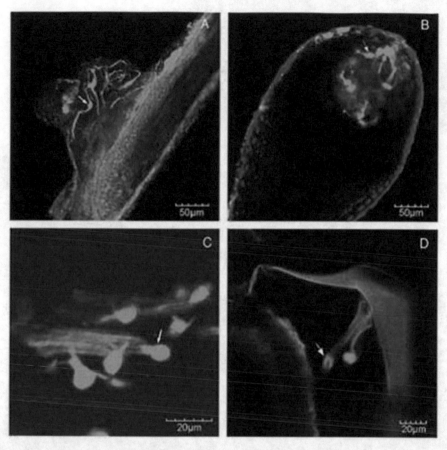

Fig. 3 (a, b) Isolated vibrissa of an ND-GFP mouse was imaged by optical sectioning in 3D using a confocal two-photon microscope (Fluoview FV1000, Olympus, Tokyo, Japan) with a high NA objective (40x/NA1.3 Oil). The figures demonstrate one optical plane image out of 3D stacks of images. BA (**a**) and DP (**b**) are in middle anagen. ND-GFP-expressing cells in the BA and DP have similar morphology (oval/round-shaped body with a diameter of approximately 5–7 μm) with long dendrite-like structures. Arrows depict typical cells in each region. The blue fluorescence is from Hoechst staining and green fluorescence is from GFP. (**c, d**) ND-GFP-expressing HAP stem cells grew out from the BA (**c**) and DP (**d**) when cultured on Gelfoam® after several days of incubation. In both images, nestin-expressing HAP stem cells from both areas have similar oval/round morphology (diameter of 5–7 μm) with long dendrite-like structures. Arrows depict typical cells in each region [14]

stem cells from BA and DP floated in culture. At day 7, spheres were formed from HAP stem cells from both regions (Fig. 4A1, A2). HAP stem cells from both regions expressed nestin, identified by immunocytochemistry (Fig. 4B1, C1). Nestin-expressing   HAP stem cells from the BA had higher sphere-forming efficiency than those from the DP (Fig. 4A3). The spheres started to differentiate after β-FGF and EGF were removed from the culture medium. Most cells were Tuj1- or GFAP-positive, suggesting that nestin-expressing HAP stem cells from both the BA and DP can differentiate into neuronal or Schwann cells (Fig. 4B2, B3, C2, C3). Nestin-expressing

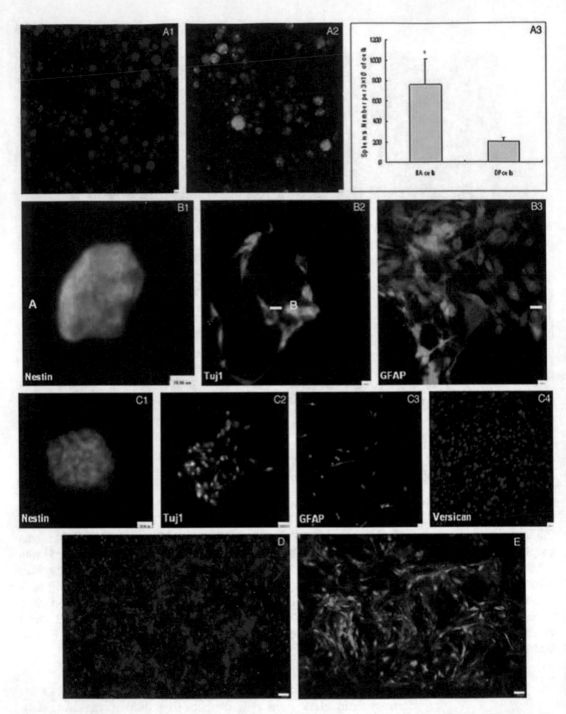

Fig. 4 Sphere-forming HAP stem cells from the BA and DP. (**A1**) Spheres formed from RFP BA HAP stem cells. (**A2**) Spheres formed from DP ND-GFP HAP stem cells. (**A3**) HAP stem cells from the BA had higher sphere-forming efficiency than those from the DP. Immunofluorescence staining showed that spheres were nestin positive (**B1**: BA sphere; **C1**: DP sphere). Tuj1- or GFAP-positive cells were detected in the spheres after β-FGF and EGF were removed from the culture medium (**B2–B3**, cells from a BA sphere; **C2–C3**, cells from a DP sphere). (**C4**) showed versican-negative staining of DP ND-GFP HAP stem cells. Green fluorescence: FITC. Red fluorescence: TRITC. Blue fluorescence: DAPI. Scales in (**A1**) and (**A2**): 50 μm. Scales in (**B1–B3, C1–C4**): 20 μm. (**D**) RFP BA HAP stem cells grown on Gelfoam® for 10 days. (**E**) DP ND-GFP HAP stem cells grown on Gelfoam® for 10 days. Scale bar: 50 μm [14]

DP HAP stem cells were versican negative (Fig. 4C4), which is a marker for DP cells. This result suggests that nestin-expressing DP HAP stem cells may not be native to the DP [14].

4.5 Three-Dimensional Culture of Nestin-Expressing BA and DP HAP stem Cells on Gelfoam®

Nestin-expressing HAP stem cells from both the BA and DP grew very well on Gelfoam®. The HAP stem cells attached to the Gelfoam® within 1 h. They grew along the grids of the Gelfoam® during the first 2 or 3 days. Later they spread into the Gelfoam® (Figs.3, 4D, E) [14].

4.6 Fate of Nestin-Expressing BA and DP HAP stem Cells Transplanted into the Injured Spinal Cord

After transplantation of Gelfoam® cultures of nestin-expressing BA or DP HAP stem cells into the injured spinal cord (including the Gelfoam®), nestin-expressing BA and DP cells were observed to be viable over 100 days post-surgery. The transplanted cells attached to the surrounding tissue very well. The surface of the transplanted area appeared smooth (Fig. 5A2, B1) [14].

When RFP-BA HAP stem cells plus Gelfoam® were transplanted into the injured spinal cord of GFP transgenic nude mice (Fig. 5A1–A8), many RFP-BA cells were found in the transplanted area (Fig. 5A3–A7). RFP BA HAP stem cells were observed migrating toward adjacent spinal cord segments (Fig. 5A7). GFP host cells and fibers also grew into the injured area (Fig. 5A5–A7). H&E staining showed connections between the transplanted HAP stem cells and the host tissue (Fig. 5A8) [14].

When ND-GFP DP Gelfoam® cultures were transplanted into the injured spinal cord (Fig. 5B1–B8), the transplanted HAP stem cells grew very well and also migrated toward adjacent spinal cord segments (Fig. 5B2, B3, B7). Some HAP stem cells had long extensions (Fig. 5B4, B5) [14].

Immunofluorescence showed many Tuj1-, Isl 1/2-, and EN1-positive cells and nerve fibers in the transplanted area of the spinal cord after BA Gelfoam® or DP Gelfoam® cultures were transplanted (Fig. 6A1–A3, B1–B3). Fewer Tuj1-, Isl 1/2-, and EN 1-positive cells or nerve fibers were observed in the Gelfoam®-only transplant group (Fig. 6C1–C3). Some cells in the transplanted area were also GFAP positive (Fig. 6A4, B4, C4) in the above three groups. The average grey value of the stained fluorescence showed that the BA Gelfoam® transplant group had significantly higher Tuj1- and EN-1-positive fluorescence grey value than the DP Gelfoam® group or Gelfoam®-only group. The Gelfoam®-only group had significantly lower Tuj1-, Isl 1/2-, and EN-1-positive fluorescence grey value than the BA Gelfoam® and DP Gelfoam® group (Fig. 6D) [14].

In the non-transplanted mice with spinal cord transection (Fig. 5C1), Tuj1 immunofluorescence staining was negative at the injured area (Fig. 5C2). Many GFAP-positive cells aggregated

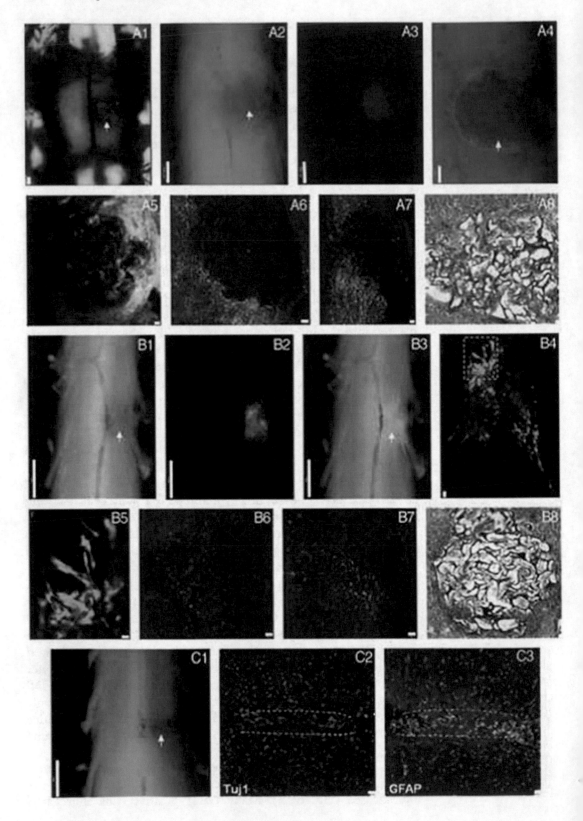

together and formed a "wall" at the injured site (Fig. 5C2). In the un-transplanted mice with spinal cord transection, glia cells formed a barrier at the spinal cord injured site. It prevented the regeneration of nerve fibers. The mice with transected spinal cords transplanted with Gelfoam® only induced less neuronal marker-positive cells in the injured site than the mice transplanted with BA HAP stem cells on Gelfoam® or DP HAP stem cells on Gelfoam® [14].

4.7 Spinal Cord Recovery After Transplantation of Nestin-Expressing BA or DP HAP Stem Cells with Gelfoam®

All the animals had paralysis of the hind limb at the same side where the spinal cord was injured. Most animals transplanted with BA HAP stem cells on Gelfoam® or DP HAP stem cells on Gelfoam® recovered plantar placing of the paw and frequent dorsal stepping within 3 days after transplantation. It took the mice transplanted with Gelfoam® 7 days to recover this behavior. It took the untransplanted group 14 days for locomotion recovery. Twenty-eight days after transplantation, the mice transplanted with HAP stem cells on Gelfoam® or DP HAP stem cells on Gelfoam® had normal locomotion. Animals walked with consistent plantar stepping, mostly coordinated, and paws in parallel at initial contact and liftoff. Though animals transplanted with Gelfoam® only also had consistent plantar stepping at day 28, they only had

---

Fig. 5 Appearance of RFP BA HAP stem cells or ND-GFP DP HAP stem cells plus Gelfoam® in the injured spinal cord. (**A1–A8**) RFP BA HAP stem cells grown on Gelfoam® were transplanted into the injured spinal cord of GFP mice. (**A1**) is day 0 after surgery, (**A2–A7**) is 5 weeks post-surgery. In panels (**A3–A7**), many RFP BA HAP stem cells were found in the transplanted area. GFP host cells and fibers grew into the injured area (**A5–A7**). In panel (**A7**), RFP BA HAP stem cells are observed migrating toward spinal cord segments. Panel (**A8**) is H&E staining of a transplanted area 7 weeks post-surgery. Many cells and fibers were found at the injured site. (**B1–B8**) A DP ND-GFP HAP stem cell Gelfoam® culture was transplanted into the injured spinal cord 8 weeks post-surgery. Transplanted cells grew very well and also migrated toward spinal cord segments (**B2, B3, B7**). Some HAP stem cells had long extensions (**B4, B5**). Panel (**B4**) is the higher magnification of panel (**B3**). Panel (**B5**) is the higher magnification of the dotted square in panel (B4). H&E staining in panel (B8) also showed cells and fibers migrating into the injured site. Arrows in panels (**A1, A2, A4, B1**, and **B3**) showed the injured and transplanted area of the spinal cord. Panels (**A1–A4, B1–B3**) were observed with the Olympus OV100. Panels (**A5–A7, B4–B7**) were observed with the IV100 Laser Scanning Microscope. Scale bar in (**A1**): 100 μm. Scale bars in (**A2**), (**A3**): 500 μm. Scale bar in (**A4**): 200 μm. Scale bars in (**A5, A8, B5, B8**): 20 μm. Scale bars in (**A6, A7, B4, B6, B7**): 50 μm. Scale bars in (**B1–B3**): 1 mm. C. Spinal cord injured on the right side without transplantation, 7 weeks post-surgery. In the injured area, many GFAP-positive cells aggregated together forming a "wall." The Tuj1 immunofluorescence staining was negative at the injured area. Dotted lines in panels 2 and 3 represent the injured area. Red fluorescence: TRITC. Blue fluorescence: DAPI. Scale bar in (**A**): 1 mm. Scale bar in (**B**), 20 μm. Panel (**C1**) shows the injured site of the spinal cord on the right side (arrow). It was observed with the OV100. Panel (**C2**) and panel (**C3**) show the immunofluorescence staining of the frozen sections of the injured spinal cord with Tuj1 and GFAP primary antibodies, observed under fluorescence microscopy. The injured area is shown by the dotted lines. It was negative for the Tuj1 immunofluorescence staining at the injured area. Many GFAP-positive cells (red fluorescence) aggregated toward the injury area to form a "glial wall." Red fluorescence was stained by TRITC-combined secondary antibody. Blue fluorescence is nuclear counterstaining with DAPI. Scale bar in (**C1**): 1 mm. Scale bars in (**C2 and C3**): 20 μm [14]

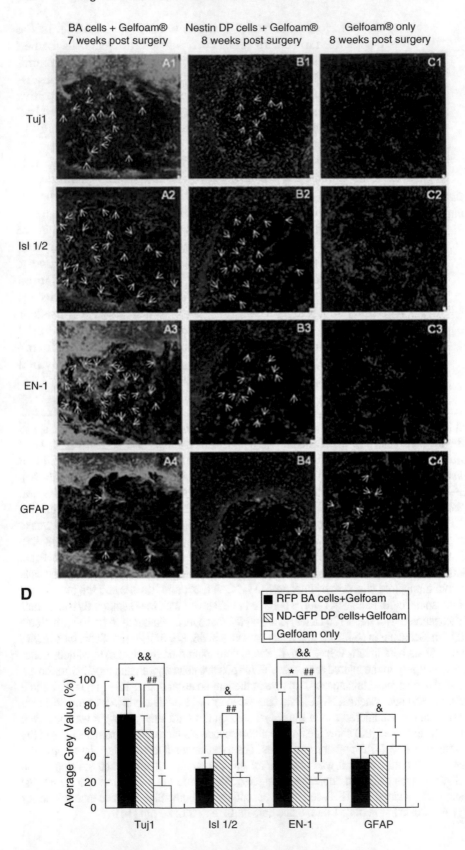

D

RFP BA cells+Gelfoam
ND-GFP DP cells+Gelfoam
Gelfoam only

some coordination or the paws rotated at initial contact and liftoff. The untransplanted mice had the least locomotion recovery. At day 28, only a few animals transplanted with Gelfoam® only walked with occasional plantar stepping. After 21 days, the BBB locomotor recovery scores of the BA Gelfoam® and the DP Gelfoam® transplant animals were similar (Fig.7). Those two groups achieved significantly greater locomotor recovery than mice transplanted with Gelfoam® only and untransplanted animals from 21 to 63 days post-surgery. The data suggest that transplanted nestin-expressing HAP stem cells from both the BA and DP have similar effects on locomotor recovery, which are better than those of Gelfoam®-only transplantation [14].

5   Notes

1. Before spraying 75% ethanol on the whisker area, it is necessary to avoid hurting the animal's eyes. Shelter the eyes with folded tissue paper.

2. The depth of whisker pad excision is determined from the size of the whisker follicles or the thickness of dermis. The depth is appropriate when the dermal papilla from the inside can be seen. Avoid cutting off the dermal papilla during the dissection.

3. HAP stem cells in anagen follicles are more active. It is better to choose anagen follicles which have an extended area between the BA and DP.

Fig. 6 Immunofluorescence staining of the transplanted area of the spinal cord. Primary antibodies used were Tuj1, Isl 1/2, EN1, and GFAP. Secondary antibodies were conjugated with FITC (green fluorescence) or TRITC (red fluorescence). Nuclei were counterstained by DAPI (blue fluorescence). Positive staining in the sections from the BA Gelfoam®-transplanted group appears green. Positive staining in the sections of the other two groups is red. In the BA or DP plus Gelfoam® transplant groups, many Tuj1-, Isl 1/2-, and EN 1-positive cells or nerve fibers (arrows) were found in the transplanted area (panels **A1–A3, B1–B3**). The Gelfoam® transplant-only group induced less neuronal marker-positive cells in the injured area than that of the BA HAP stem cells + Gelfoam® or DP HAP stem cells + Gelfoam® transplant groups. However, the Gelfoam®-only group had more GFAP-positive cells than that in the BA or DP Gelfoam® groups. Scale bar: 20 μm. Glial cells could form a barrier in the spinal cord injured area. The barrier prevented the regeneration of nerve fibers. In the Gelfoam®-only group, some neuronal-like cells appeared in the transplanted area (**C1, C3**). Those cells may come from the surrounding spinal cord tissue. More neurons or nerve fibers appeared in the BA or DP Gelfoam® groups, which stimulated the recovery of spinal cord injury much more than Gelfoam® only. The neurons in the two former groups could come from the implanted BA HAP stem cells or DP HAP stem cells or migrate from the surrounding tissue. (**D**) Average grey value of the fluorescence in the transplanted area of spinal cord. The BA Gelfoam® group had a significantly higher Tuj1- and EN-1-positive fluorescence grey value than the DP Gelfoam® group and Gelfoam®-only group. The Gelfoam®-only group had significantly lower Tuj1-, Isl 1/2-, and EN-1-positive fluorescence grey value than that in the BA Gelfoam® and DP Gelfoam® groups, except for the GFAP-positive fluorescence grey value. *, BA Gelfoam® group vs. DP Gelfoam® group, $p < 0.05$; &, BA Gelfoam® group vs. Gelfoam®-only group, $p < 0.05$; &&, $p < 0.01$; ##, ND-GFP DP cells + Gelfoam® group vs. Gelfoam®-only group, $p < 0.01$ [14]

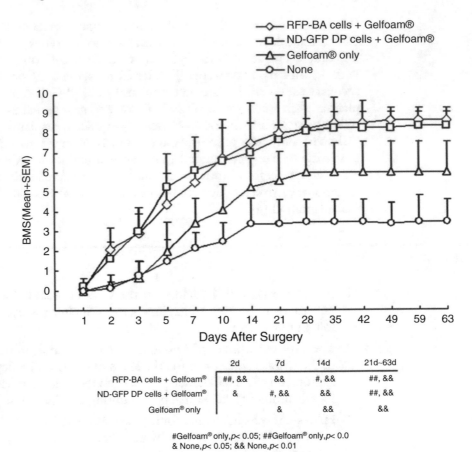

Fig. 7 Mean Basso mouse scale (BMS) for locomotion (BMS) scores. The locomotor recovery pattern of the BA Gelfoam® and the DP Gelfoam® animals was similar. Those two groups achieved significantly greater locomotor recovery than the Gelfoam®-only and non-transplanted control animals [14]

4. There are two blood-filled areas of each whisker follicle which is sealed by an outer connective capsule. Cut the follicle into half just between the two blood-filled areas.

5. After tearing off the outer connective capsule, the blood-filled areas are stirred. The BA is visible at the insertion point of the arrector pili muscle. The DP is a small soft "ball" at the lowest site of the follicle. It is recommended to separate the DP with a 31G insulin syringe needle and transfer the DP with the tip of a 1 ml syringe needle for use.

6. The plates for BA or DP adhesive culture need to be incubated with primary culture medium at 37 °C, 95% CO_2, 5% O_2, at the beginning of the dissection. The medium is removed entirely just before use. During placing of the BA or DP into the plate, forceps or needles may bring medium into the plate. It is important to remove the extra medium around each BA or DP to avoid floating of the tissue.

7. During cell injection into the Gelfoam®, it is recommended to inject cells in the deeper site at the beginning and then in the middle and superficial site of the Gelfoam® sponge. One 3 mm³ Gelfoam® sponge has six injection points. All the processes are performed under a stereomicroscope.

8. Normally the mouse has 13 pairs of ribs. It is easier to determine the 13th thoracic (T13) vertebra according to the last rib of mouse. The 8 vertebra is three segments cranial to the T11 vertebra, which is the highest in the mouse spine.

9. The needle needs to be placed vertically during the semi-transection of the spinal cord. It is necessary to observe the blood vessels on the spinal cord carefully under a stereomicroscope. The median posterior spinal vessel lies longitudinally at the center of the spinal cord, which must not be destroyed. This blood vessel could be a marker to identify the right and left halves of the spinal cord.

10. The Gelfoam®-HAP stem cell complex should be placed in a tissue culture dish with a little PBS before dividing it. Place a sharp double-edged blade 1 mm parallel to one side of the Gelfoam® and make a quick cut vertically from the surface to the bottom under a stereomicroscope as quickly as possible.

Acknowledgement

The work reviewed in the present chapter was partially-supported by the National Natural Science Foundation of China (no. 81571211) and the Natural Science Foundation of Shanghai, China (no. 14ZR1449300).

References

1. Li L, Mignone J, Yang M, Matic M, Penman S, Enikolopov G, Hoffman RM (2003) Nestin expression in hair follicle sheath progenitor cells. Proc Natl Acad Sci U S A 100:9658–9661

2. Sieber-Blum M, Grim M, Hu YF, Szeder V (2004) Pluripotent neural crest stem cells in the adult hair follicle. Dev Dyn 231:258–269

3. Sieber-Blum M, Schnell L, Grim M, Hu YF, Schneider R, Schwab ME (2006) Characterization of epidermal neural crest stem cell (EPI-NCSC) grafts in the lesioned spinal cord. Mol Cell Neurosci 32:67–81

4. Amoh Y, Li L, Katsuoka K, Penman S, Hoffman RM (2005) Multipotent nestin-positive, keratin-negative hair-follicle-bulge stem cells can form neurons. Proc Natl Acad Sci U S A 102:5530–5534

5. Biernaskie J, Paris M, Morozova O, Fagan BM, Marra M, Pevny L, Miller FD (2009) SKPs derive from hair follicle precursors and exhibit properties of adult dermal stem cells. Cell Stem Cell 5:610–623

6. Hoffman RM (2006) The pluripotency of hair follicle stem cells. Cell Cycle 5:232–233

7. Amoh Y, Kanoh M, Niiyama S, Kawahara K, Satoh Y, Katsuoka K, Hoffman RM (2009) Human and mouse hair follicles contain both multipotent and monopotent stem cells. Cell Cycle 8:176–177

8. Yashiro M, Mii S, Aki R, Hamada Y, Arakawa N, Kawahara K, Hoffman RM, Amoh Y (2015) From hair to heart: nestin-expressing hair-follicle-associated pluripotent (HAP) stem cells differentiate to beating cardiac muscle cells. Cell Cycle 14:2362–2366

9. Amoh Y, Li L, Campillo R, Kawahara K, Katsuoka K, Penman S, Hoffman RM (2005) Implanted hair follicle stem cells form Schwann cells that support repair of severed peripheral nerves. Proc Natl Acad Sci U S A 102:17734–17738

10. Amoh Y, Li L, Katsuoka K, Hoffman RM (2008) Multipotent hair follicle stem cells promote repair of spinal cord injury and recovery of walking function. Cell Cycle 7:1865–1869

11. Yu H, Fang D, Kumar SM, Li L, Nguyen TK, Acs G et al (2006) Isolation of a novel population of multipotent adult stem cells from human hair follicles. Am J Pathol 168:1879–1888

12. Yu H, Kumar SM, Kossenkov AV, Showe L, Xu X (2010) Stem cells with neural crest character-istics derived from the bulge region of cultured human hair follicles. J Invest Derm 130: 1227–1236

13. Amoh Y, Kanoh M, Niiyama S, Hamada Y, Kawahara K, Sato Y et al (2009) Human hair follicle pluripotent stem (hfPS) cells promote regeneration of peripheral-nerve injury: an advantageous alternative to ES and iPS cells. J Cell Biochem 107:1016–1020

14. Liu F, Uchugonova A, Kimura H, Zhang C, Zhao M, Zhang L, Koenig K, Duong J, Aki R, Saito N, Mii S, Amoh Y, Katsuoka K, Hoffman RM (2011) The bulge area is the major hair follicle source of nestin-expressing pluripotent stem cells which can repair the spinal cord compared to the dermal papilla. Cell Cycle 10:830–839

Chapter 16

Nerve Growth and Interaction in Gelfoam® Histoculture: A Nervous System Organoid

Robert M. Hoffman, Sumiyuki Mii, Jennifer Duong, and Yasuyuki Amoh

Abstract

Nestin-expressing hair follicle-associated pluripotent (HAP) stem cells reside mainly in the bulge area (BA) of the hair follicle but also in the dermal papilla (DP). The BA appears to be origin of HAP stem cells. Long-term Gelfoam® histoculture was established of whiskers isolated from transgenic mice, in which there is nestin-driven green fluorescent protein (ND-GFP). HAP stem cells trafficked from the BA toward the DP area and extensively grew out onto Gelfoam® forming nerve-like structures. These fibers express the neuron marker β-III tubulin-positive fibers and consisted of ND-GFP-expressing cells and extended up to 500 mm from the whisker nerve stump in Gelfoam® histoculture. The growing fibers had growth cones on their tips expressing F-actin indicating that the fibers were growing axons. HAP stem cell proliferation resulted in elongation of the follicle nerve and interaction with other nerves in 3D Gelfoam® histoculture, including the sciatic nerve, trigeminal nerve, and trigeminal nerve ganglion.

Key words Nestin, Green fluorescent protein, Hair follicle, Stem cells, Sensory nerve, Sciatic nerve, Trigeminal nerve, Nerve interaction, Gelfoam® histoculture

1 Introduction

Nestin, which marks progenitor cells of the CNS, is also expressed in cells which occupy the permanent upper hair follicle immediately below the sebaceous glands in the hair follicle bulge. These cells are marked by green fluorescent protein (GFP) fluorescence in transgenic mice, in which the nestin promoter drives GFP (nestin-driven green fluorescent protein [ND-GFP]) [1, 2]. ND-GFP stem cells isolated from the hair follicle bulge area can differentiate into neurons, glia, keratinocytes, smooth muscle cells, heart muscle, and melanocytes in vitro. These multipotent ND-GFP stem cells are positive for the stem cell marker CD34, as well as keratin-15 [3]. These cells are termed hair follicle-associated-pluripotent (HAP) stem cells.

When the HAP stem cells were implanted into the gap region of the severed sciatic nerve of mice, they greatly enhanced the rate of nerve regeneration and the restoration of nerve function. The stem cells transdifferentiated largely into Schwann cells, which are

Robert M. Hoffman (ed.), *3D Sponge-Matrix Histoculture: Methods and Protocols*, Methods in Molecular Biology, vol. 1760, https://doi.org/10.1007/978-1-4939-7745-1_16, © Springer Science+Business Media, LLC, part of Springer Nature 2018

known to support neuron regrowth. The transplanted mice recovered the ability to walk normally [4].

HAP stem cells were also transplanted to the injured thoracic spinal cord of C57BL/6 immunocompetent mice. Most of the transplanted cells also differentiated into Schwann cells that apparently facilitated repair of the spinal cord. The rejoined spinal cord reestablished extensive hind-limb locomotor performance [5].

In the human hair follicle dissected from the scalp, the cells immediately below the sebaceous glands, just above the bulge area, were observed to be nestin positive and therefore HAP stem cells. The human HAP stem cells could differentiate into β3-tubulin-positive neurons, S-100-positive and GFAP-positive glial cells, K15-positive keratinocytes, and SMA-positive smooth muscle cells, similar to nestin-expressing hair follicles of the mouse [6–8].

A subsequent study in the mouse showed that HAP stem cells were present in both the bulge area (BA) and dermal papilla (DP). The ability of HAP stem cells from each region were compared to repair spinal cord injury. In the whiskers of ND-GFP transgenic mice, the DP was found to contain HAP stem cells only in early and middle anagen. The BA had HAP stem cells throughout the hair cycle and to a greater extent than the DP. HAP stem cells from both areas differentiated into neuronal cells at high frequency in vitro. Both DP and BA HAP stem cells differentiated into neuronal and Schwann cells after transplantation to the injured spinal cord and enhanced injury repair and locomotor recovery [9].

The HAP stem cells from both the BA and DP have a small body diameter of approximately 7 μm with long protrusions shown by confocal imaging. The BA is the source of the HAP stem cells of the hair follicle and they migrate from the BA to the DP as well as into the surrounding skin tissues including the epidermis, during wound healing. These results suggest that the BA may be the source of the stem cells of the skin itself [10].

Yu et al. [7] cultured a population of human nestin-expressing HAP stem cells. The human bulge area HAP stem cells proliferated as spheres, were capable of self-renewal, and differentiated into multiple lineages including myogenic, melanocytic, and neuronal cell lineages in vitro. Human HAP stem cells were transplanted in the severed sciatic nerve of the mouse which differentiated into glial fibrillary acidic protein (GFAP)-positive Schwann cells and affected the functional recovery of the nerve [8].

2 Materials

2.1 Animal and Whisker Dissection

1. Animals: Nestin-driven GFP (ND-GFP) transgenic mice; non-transgenic C57BL/6 mice; red fluorescent protein (RFP) transgenic mice.

2. Dissection or surgery equipment: Stereomicroscope, surgical knife, fine forceps, fine scissors, retractor for mice, syringe needles (10 ml, 5 ml, 1 ml syringe), 31G insulin syringe.

3. Tissue and cell culture plates or flasks: Tissue culture dishes (60 mm), collagen I-coated 12-well and 48-well plates, collagen I-coated 25 cm^2 flasks.

4. Culture medium: Dulbecco's modified Eagle's medium (DMEM)/F12 (GIBCO, Grand Island, NY); 10% fetal bovine serum (FBS) (GIBCO); 1% N2 (GIBCO); 2% B27 (GIBCO); 200 mM L-glutamine (GIBCO); 0.025% ITS +3 (Sigma Aldrich, St. Louis, MD); 20 ng/ml epithelial growth factor; 20 ng/ml basic fibroblast growth factor (Invitrogen, Carlsbad, CA). Subculture medium is similar to the culture medium but serum free or hair follicle growth medium containing, per 100 ml, William's E medium [11], 1 ml L-glutamine, 20 µl of a hydrocortisone mixture (1 mg hydrocortisone powder, 1 ml 100% ethanol, 19 ml William's E medium), 100 µl insulin mixture (100 mg bovine insulin, 100 µl sterile glacial acetic acid, 10 ml sterile pure water), and 1 ml 1% penicillin/streptomycin (Gibco/BRL).

2.2 Gelfoam® Preparation

1. Sterile collagen sponges: Gelfoam® Absorbable Collagen Sponge USP, 12-7 mm format (Pfizer, NDC: 0009-0315-08).

2. Gelfoam® trimming aid: Double-edged blade.

3. Sterile metal forceps or tweezers.

4. Sterile metal scissors.

5. Sterile flat weighing metal spatula.

6. 6-Well culture plates.

7. Water-jacketed CO_2 incubator, set at 37 °C, 5% CO_2, ≥90% humidity.

8. Water bath.

9. 70% ethanol or isopropyl alcohol solution.

2.3 Imaging

1. Binocular microscope (MZ6) (Leica).

2. FV1000 confocal microscope (Olympus, Tokyo, Japan) or equivalent.

3. OV100 Small Animal Imaging System (Olympus, Tokyo, Japan) or equivalent.

4. IV100 Laser Scanning Microscope (Olympus, Tokyo, Japan) or equivalent.

2.4 Immuno-fluorescence Reagents

1. Paraformaldehyde.

2. Tissue-freezing medium (Triangle Biomedical Science, Durham, NC).

3. Liquid nitrogen.

4. CM1850 cryostat (Leica).

5. Phosphate-buffered saline (PBS).

6. Hematoxilin and eosin (H&E).

7. 5% normal goat serum.

8. DAPI (1:48,000; Invitrogen).

9. Fluoromount (Sigma, St. Louis, MO).

2.4.1 *Primary Antibodies Used are:*

1. Anti-β III tubulin mAb (mouse, 1:100) (Santa Cruz Biotech, Dallas, TX).

2. Anti-glial fibrillary acidic protein (GFAP) mAb (mouse, 1:250) (BD Pharmingen, San Jose, CA).

3. Anti-S100 mAb (mouse, 1:200) (Millipore, Billerica, MA).

4. Anti-p75NTR mAb (rabbit, 1:3,200) (Cell Signaling, Danvers, MA).

5. Anti-TrkA mAb (rabbit, 1:50) (Santa Cruz Biotech).

6. Anti-TrkB mAb (rabbit, 1:50) (Santa Cruz Biotech).

2.4.2 *Secondary Antibodies Used are:*

1. Goat antimouse IgG (H+L) Alexa Fluor® 555 (1:1,000; Cell Signaling).

2. Goat anti-rabbit IgG (H+L) Alexa Fluor® 555 (1:1,000; Cell Signaling).

3. Alexa Fluor® 647 Phalloidin (1:40; Invitrogen, Grand Island, NY).

3 Methods

3.1 Isolation of ND-GFP Transgenic Mice Vibrissa (Whisker Hair Follicles) from Whisker Pads

1. Trim and sterilize whiskers and whisker pads with 70% isopropyl alcohol.

2. Isolate whisker pads from ND-GFP mice under anesthesia with surgical scissors and wash with alcohol, and then phosphate-buffered saline three times.

3. Separate single vibrissa using a binocular microscope from the whisker pads with fine insulin needles.

4. Place the individual whisker follicles on pre-incubated Gelfoam® in medium for subsequent histoculture (please see below) [2].

3.2 Isolation of Trigeminal and Sciatic Nerves and the Trigeminal Nerve Ganglion

1. In order to excise the trigeminal nerves, sterilize whisker pads from RFP transgenic mice, with 70% isopropyl alcohol. Expose the trigeminal nerve from the infraorbital foramen in each vibrissa, on the inside of the whisker pad, and remove [12].

2. In order to isolate the sciatic nerve, make a skin incision in the medial side of the thigh of RFP transgenic mice.

3. Expose the nerve between the short adductor muscle and long adductor muscle [12].

4. In order to isolate the trigeminal nerve ganglion, make a skin incision on top of the head and the skull and open with a drill.

5. Expose the trigeminal nerve ganglion after the skull base is removed.

6. Using the binocular microscope, excise the trigeminal nerve ganglion with fine forceps [12].

3.3 Gelfoam®
Histoculture
and Growth Medium

1. Place the vibrissae hair follicles, with or without capsules and containing their sensory nerve stumps, as well as trigeminal and sciatic nerves and the trigeminal nerve ganglion, on sterile Gelfoam®, hydrated in cell culture medium [2].

2. Arrange the whisker sensory nerve stump and the trigeminal nerve or sciatic nerve with stumps opposed to each other.

3. Arrange the vibrissae hair follicle containing the sensory nerve stump and the trigeminal nerve ganglion with the nerve stump opposed to the ganglion.

4. Use DMEM-F12 culture medium containing B-27 (2.5%), N2 (1%), and 1% penicillin and streptomycin.

5. Incubate the culture at 37 °C, 5% CO_2, 100% humidity.

6. Change the medium every other day [2, 13].

3.4 Confocal
Microscopy

1. Prior to imaging, remove the growth medium from the dish.

2. Use a confocal microscope for two (x, y)-dimensional high-resolution imaging of the histocultured whiskers [9, 10] with laser excitation at 470 nm for GFP.

3. Obtain images using the ×20/0.50 UPlan FLN and ×10/0/30 Plan NEOFLUAR objectives [2].

3.5 Immuno-
fluorescence Methods

3.5.1 Preparation
of Slides

1. Fix tissues in pre-cooled 4% paraformaldehyde at room temperature (RT) for 2 h.

2. Embed tissues in tissue-freezing medium and freeze in nitrogen for 10 min at −80 °C.

3. Prepare frozen sections of 7–10 mm thickness with a CM1850 cryostat (Leica).

4. Wash the frozen sections with PBS three times.

5. Process frozen sections for hematoxilin and eosin (H&E) staining.

3.5.2 Stain for
Immunofluoresence
as Follows

1. Apply 5% normal goat serum at RT for 1 h.

2. Apply the primary antibodies at RT for 2 h.

3. Apply the secondary antibodies at RT in the dark for 1 h.

4. For detection of F-action, incubate at RT in the dark for 1 h.

5. For DAPI, incubate at RT in the dark for 3 min.

6. Mount slides with Fluoromount and observe under confocal laser scanning microscopy.

4   Results and Discussion

4.1   HAP Stem Cells Migrate from the BA to the DP

A typical isolated whisker from ND-GFP mice can be seen in Fig. 1 where the bright-field image shows the BA and DP. Figure 2 shows a series of time-course images depicting a whisker and the trafficking HAP stem cells within it over a period of 11 days in histoculture on Gelfoam®.

Figure 3 shows another whisker at histoculture day 12. The BA does not contain many HAP stem cells. However, many HAP stem cells are present at the side of the DP. The HAP stem cells appear to gather and orient in a bundle-like structure at the side of the DP. Extensive HAP stem cell outgrowth that also appears to be oriented in bundles occurs in the Gelfoam® [2].

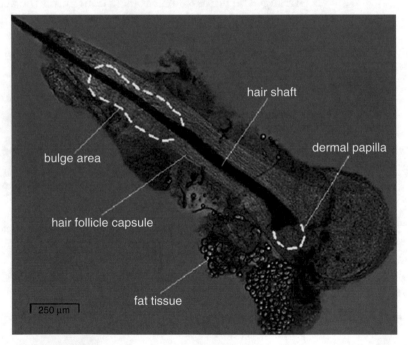

Fig. 1 Image of a vibrissa follicle excised from the whisker pad of a transgenic mouse expressing ND-GFP. White dashed lines depict the bulge area and dermal papilla [2]

Fig. 2 (continued) The HAP stem cells appear to gather and orient at the side of the DP (white box) in a bundle-like structure (yellow arrows). **B1** is shown at a higher magnification in **B2**. Single HAP stem cells (white arrows) are visible throughout the whisker, but more are present at the DP end. **B3** shows the DP area at a higher magnification. Images were observed using the FV1000. Scale bar in **B1**, 500 μm; scale bars in **B2** and **B3**, 100 μm. **C1–C3** show images of HAP stem cells in the whisker in histoculture on day 11. **C1** shows the whole whisker with BA (red box) and DP (blue box). HAP stem cells (white arrows) are oriented to the side (white box) of the DP area, as they gather in bundle-like structures (yellow arrows). The HAP stem cells are shown at a higher magnification in **C2**. HAP stem cell outgrowth (purple box) into the Gelfoam® in **C3** (not seen in **C1**) is visible because the stem cells are located in a different layer on the Gelfoam®. The white dashed lines depict the whisker follicle perimeter. Images were obtained with the FV1000. Scale bar in **C1**, 500 μm; scale bars in **C2** and **C3**, 100 μm [2]

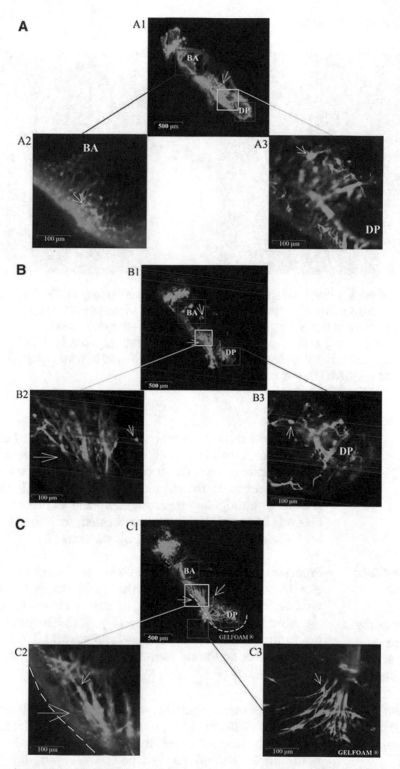

Fig. 2 A1–A3 are images of HAP stem cells in whisker histoculture on day 0. **A1** shows the whole whisker, with labeled BA (red box) and DP areas (blue box). The area adjacent to the DP (white box) is shown at a higher magnification in **A3**. **A2** shows HAP stem cells in the BA at a higher magnification. Many HAP stem cells (white arrows) are present throughout the whisker. Images were observed using the Olympus FV1000 confocal microscope. Scale bar in **A1**, 500 μm; scale bars in **A2** and **A3**, 100 μm. **B1–B3** show images of HAP stem cells in whisker histoculture on day 5. **B1** shows the whole whisker with BA (red box) and DP (blue box).

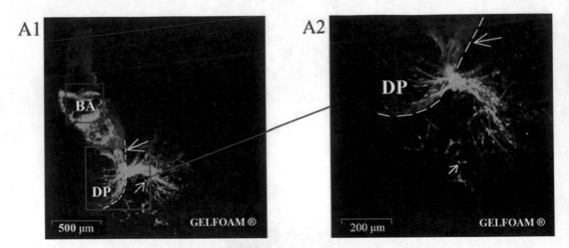

Fig. 3 A1–A2 are views of single whisker in histoculture on Gelfoam® on day 12. The BA (red box) does not contain many HAP stem cells (white arrows), while many stem cells are present at the side of the DP (blue box). The DP area is shown at a higher magnification in **A2**. The HAP stem cells appear to gather and orient in a bundle-like structure (yellow arrows) at the side of the DP. Stem cell outgrowth is visible on Gelfoam®. The white dashed lines depict the hair follicle perimeter. Images were obtained with the Olympus FV1000 Imaging System. The scale bar in **A1**, 500 μm; scale bar in **A2**, 200 μm [2]

Figure 4 shows whiskers at histoculture day 11. HAP stem cells have gathered into bundle-like structures at the side of the hair follicle, consistent with the location at which the whisker sensory nerve is present in the diagram in Fig. 4A1. HAP stem cell outgrowth in bundle-like structures into the Gelfoam® can be seen. Five sets of experiments were performed, with four to seven whisker follicles in each set, totaling 32 whiskers [2].

4.2 HAP Stem Cells Extended Their Processes and Migrated Toward the Whisker Sensory Nerve Stump in 3D Gelfoam® Histoculture

Processes from the HAP stem cells in the vibrissa hair follicle bulge area began to extend toward the nerve stump by day 4 in 3D Gelfoam® histoculture (Fig. 5). By day 9, HAP stem cells reached the whisker sensory nerve stump (Fig. 2B). The processes extending from the HAP stem cells co-expressed β-III tubulin, $p75^{NTR}$, and TrkB but no longer expressed S100 (Fig. 5) [12].

4.3 HAP Stem Cells Proliferated at the Whisker Sensory Nerve Stump and Formed Cord-Like Structures in 3D Gelfoam® Histoculture

The whisker sensory nerve stump became enriched with HAP stem cells in 3D Gelfoam® histoculture. Higher magnification images of the whisker sensory nerve stump demonstrated the presence of both spindle- and spherical-shaped cells (Fig. 6, day 8). The HAP stem cells formed cord-like structures extending further into the Gelfoam® from the sensory nerve stump (Fig. 6, day 16). The spindle-shaped cells highly expressed ND-GFP. However, only some of the round-shaped cells expressed ND-GFP, which eventually was extinguished (Fig. 6). The cord-like structures continued to extend at day 21 in Gelfoam® culture [12].

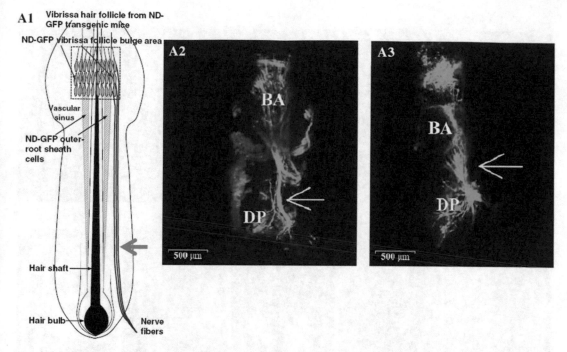

Fig. 4 Hair follicle diagram (**A1**) [3] and two hair follicles in histoculture on Gelfoam® on day 11 (**A2** and **A3**). HAP stem cells have gathered into bundle-like structures (yellow arrows) at the side of the hair follicle, consistent with the location in which the whisker sensory nerve is present (blue arrow) (**A1**). HAP stem cells also grow out onto the Gelfoam® [2]

4.4 The Whisker Sensory Nerve Interacted with the Trigeminal Nerve, Mediated by HAP Stem Cells, in 3D Gelfoam® Histoculture

The vibrissa follicle obtained from the ND-GFP transgenic mice, containing its sensory nerve stump, was co-cultured on Gelfoam® along with the trigeminal nerve from RFP transgenic mice (Fig. 7). The GFP-expressing cord-like structures of the whisker sensory nerve intermingled with the ends of the co-cultured RFP-expressing trigeminal nerve by day 12 of 3D Gelfoam® histoculture (Fig. 7). The HAP stem cells proliferated at the whisker sensory nerve stump, forming radial cords which extended the nerve. By day 10, the thickest cord extended and intermingled with the RFP-expressing trigeminal nerve. Two types of cells were observed in the histocultured whisker nerve intermingling with the trigeminal nerve; one was spindle-shaped highly expressing ND-GFP throughout the nerve, and the other was round shaped with less ND-GFP expression (Fig. 7) [12].

4.5 The Whisker Sensory Nerve Intermingled with the Sciatic Nerve, Mediated by HAP Stem Cells, in 3D Gelfoam® Histoculture

Figure 7 shows a time course of the interaction of the GFP-expressing whisker sensory nerve and a sciatic nerve from an RFP transgenic mouse in 3D Gelfoam® histoculture. By day 9 of 3D Gelfoam® histoculture, the HAP stem cells migrated from the whisker sensory nerve stump of the vibrissa hair follicle and invaded deeply into the RFP-expressing sciatic nerve (Fig. 7) [12].

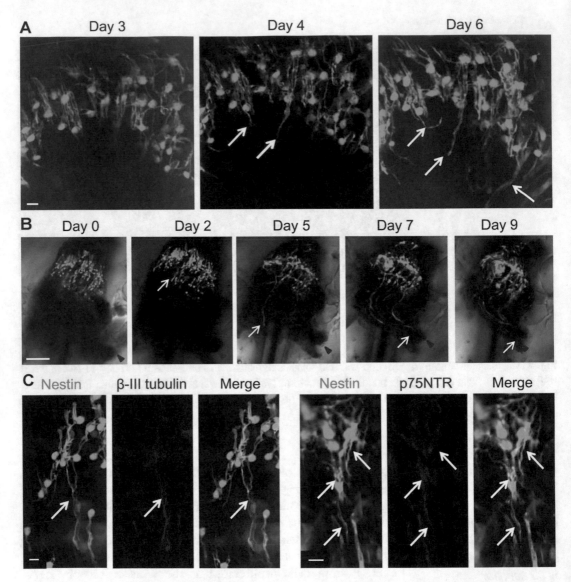

Fig. 5 HAP stem cells extended their processes and migrated toward the whisker sensory nerve stump in histoculture. (**a**) Vibrissa hair follicle without its capsule in 3D Gelfoam® histoculture. The processes (white arrows) from the HAP stem cells appeared and began to extend by day 4 of 3D Gelfoam® histoculture. White bar: 10 mm. (**b**) The processes (white arrow) extended from the HAP stem cells by day 2 in 3D Gelfoam® histoculture. By day 9, HAP stem cells (white arrow) reached the whisker sensory nerve stump (red arrowhead). White bar: 100 mm. (**c**) Vibrissa hair follicles cultured in 3D Gelfoam® histoculture for 1 week. The processes (white arrows) extending from the HAP stem cells co-expressed β-III tubulin and p75NTR. White bar: 10 mm [12]

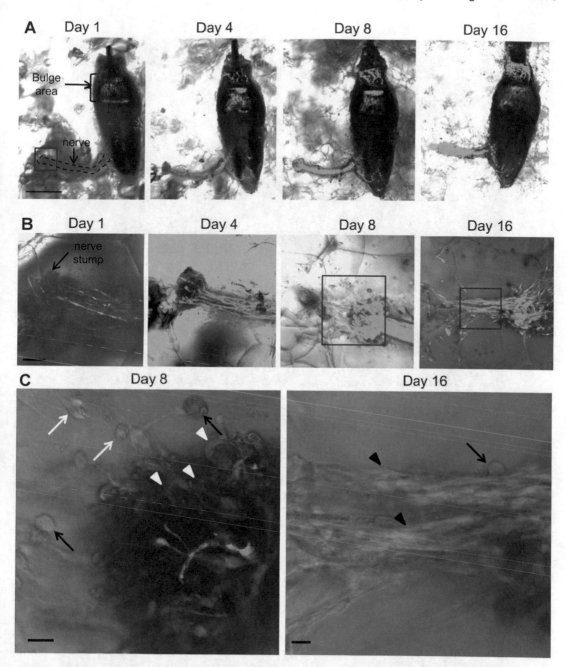

Fig. 6 HAP stem cells proliferated at the whisker sensory nerve stump and formed cord-like structures in histoculture. (**a**) Time-course imaging of HAP stem cell trafficking in a histocultured vibrissa hair follicle from an ND-GFP mouse. The nerve stump became enriched with HAP stem cells by day 8 in 3D Gelfoam® histoculture. Bar: 500 mm. (**b**) Magnified images of the nerve stump. By day 8, the HAP stem cells proliferated at the nerve stump and migrated into the Gelfoam®. The HAP stem cells formed cord-like structures extending 500 mm from the whisker sensory nerve stump by day 16. Bar: 100 mm. (**c**) Higher magnification images of the area inside the boxes in (**b**). By day 8, the HAP stem cells grew out from the nerve stump (white arrowheads). Some round-shaped cells expressed nestin (white arrows) and some did not (black arrows). By day 16, the cord-like structure contained spindle-shaped cells expressing ND-GFP (black arrowheads). In addition, there were round-shaped cells without nestin expression (black arrow). Bar: 10 mm [12]

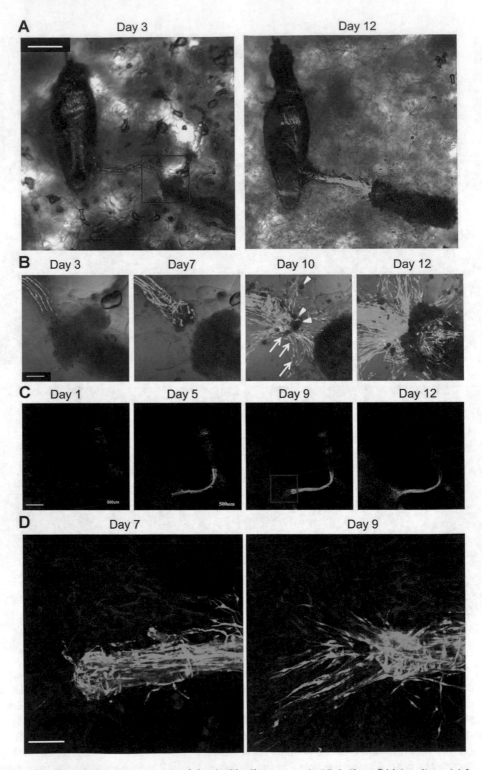

Fig. 7 The extending whisker sensory nerve joined with other nerves in 3D Gelfoam® histoculture. (**a**) A whisker sensory nerve of a vibrissa hair follicle from an ND-GFP transgenic mouse was placed on Gelfoam® next to the trigeminal nerve from an RFP transgenic mouse. By day 12, in 3D Gelfoam® histoculture the whisker sensory nerve was enriched with HAP stem cells and intermingled with the RFP-expressing trigeminal nerve.

4.6 HAP Stem Cells Expressed Neuronal Markers and Nestin-Negative Round-Shaped Cells Expressed Glial Markers in the Extending Whisker Sensory Nerve Interacting with the Trigeminal Nerve in 3D Gelfoam® Histoculture

Two whisker sensory nerves were arranged on either side of a trigeminal nerve on Gelfoam® and cultured. The nerves intermingled within 1 month of histoculture (Fig. 8). The HAP stem cells in the whisker sensory nerve, intermingling with the trigeminal nerve, co-expressed p75^{NTR}, TrkB, and β-III tubulin (Fig. 8). High-magnification images show that spindle-shaped cells expressing ND-GFP in the whisker nerve co-expressed p75^{NTR}, TrkB, and β-III tubulin (Fig. 8). Most of the many round-shaped cells expressed GFAP, but not nestin (Fig. 8) [12].

4.7 The Growing Whisker Sensory Nerve Expressed β-III Tubulin and Phalloidin-Positive F-Actin

By day 15 in 3D Gelfoam® histoculture (Fig. 9), many β-III tubulin-positive fibers extended from the nerve stump. The fibers consisted of HAP stem cells (Fig. 9). By 28 days of 3D Gelfoam® histoculture, the β-III tubulin-positive fibers extended widely and radially around the hair follicle sensory nerve (Fig. 9). The tips of the β-III tubulin-positive fibers expressed phalloidin-positive F-actin suggesting that the β-III tubulin-positive fibers were axons growing from the whisker sensory nerve stump (Fig. 9) [12].

4.8 The Extending Hair Follicle Sensory Nerve Interacted with the Trigeminal Nerve Ganglion in 3D Gelfoam® Histoculture

A vibrissa hair follicle, containing its sensory nerve stump, was co-cultured on Gelfoam® with a trigeminal nerve ganglion, both isolated from an ND-GFP transgenic mouse (Fig. 10). The extending GFP-expressing cord-like structures of the whisker sensory nerve intermingled extensively with the co-cultured trigeminal nerve ganglion by day 41 of 3D Gelfoam® histoculture (Fig. 10). Many β-III tubulin-positive fibers extended from both the trigeminal nerve ganglion and the hair follicle sensory nerve. The fibers consisted of HAP stem cells (Fig. 10). The β-III tubulin-positive fibers extending from the nerve stump of the whisker spread widely like a fan and extended toward the trigeminal nerve ganglion. In long-term Gelfoam® histoculture, there was a thick bundle of fibers linking the trigeminal nerve ganglion and the whisker sensory nerve stump (Fig. 10) [12] (*see* **Note 1**).

Fig. 7 (continued) White bar: 500 mm. (**b**) Magnified images of the area inside the box in (**a**) show that the HAP stem cells proliferated at the whisker sensory nerve stump, forming radial cords which extended the nerve. By day 10, the thickest cord extended and intermingled with the RFP-expressing trigeminal nerve. By day 12, the two nerves intermingled. Fibers containing HAP stem cells migrated into the RFP-expressing trigeminal nerve. There were two types of the growing whisker sensory nerve; one was spindle shaped, expressing ND-GFP throughout (white arrows), and the other was round shaped with less nestin expression (white arrowheads). Both cell types were present at the junction of the two nerves. White bar: 100 mm. (**c**) Time-course imaging of the intermingling of the whisker sensory nerve containing ND-GFP cells and the sciatic nerve from an RFP mouse in Gelfoam® histoculture. White bar: 500 mm. (**d**) High-magnification images show that HAP stem cells migrated from the nerve stump of the vibrissa and invaded deeply into the RFP-expressing sciatic nerve shown at days 7 and 9 of 3D Gelfoam® histoculture. White bar: 100 mm [12]

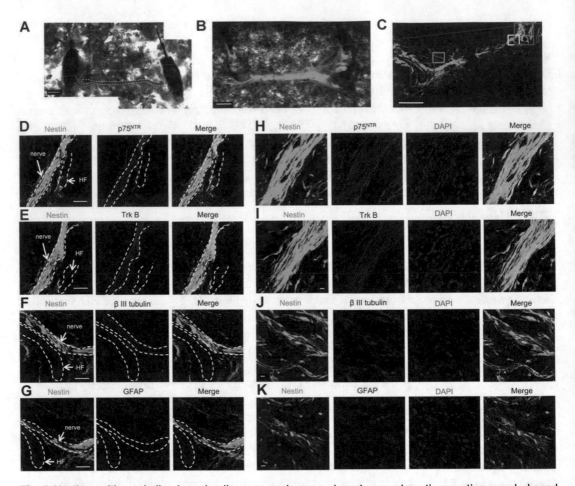

Fig. 8 Nestin-positive spindle-shaped cells expressed neuronal markers and nestin-negative round-shaped cells expressed a glial marker in the growing whisker sensory nerve intermingling with the trigeminal nerve in 3D histoculture. (**a**) Two whisker sensory nerves (enclosed by green dashed lines) were arranged on either side of a trigeminal nerve (enclosed by yellow dashed lines) on Gelfoam®. White bar: 500 mm. (**b**) The nerves intermingled by 1 month of Gelfoam® histoculture. White bar: 500 mm. (**c**) One section from (**b**). White bar: 500 mm. (**d–f**) Magnified images from the blue box in (**c**). (**f**) contains magnified images from the red box in (**c**). The ND-GFP-expressing cells in the nerve co-expressed p75^{NTR}, TrkB, and β-III tubulin. White bar: 100 mm. (**g**) A magnified image from the red box in (**c**). There were many round-shaped cells proliferating around the hair follicle and the nerve. Most of these cells expressed GFAP but not ND-GFP. White bar: 100 mm. (**h, i**) High-magnification images of the area inside the yellow box in (**c**) show that the spindle-shaped cells, expressing ND-GFP in the nerve, co-expressed p75^{NTR} and TrkB. White bar: 10 mm. (**j**) Magnified image from the area inside the green box in (**c**). The spindle-shaped cells expressing nestin-GFP co-expressed β-III tubulin. White bar: 10 mm. (**k**) Magnified image of the area inside the green box in (**c**). The nestin-negative round-shaped cells expressed GFAP. White bar: 10 mm. HF: hair follicle [12]

Fig. 9 (continued) images of β-III tubulin (red) demonstrated that many β-III tubulin-positive fibers extended from the nerve stump. White bar: 500 mm. (**c**) Merged images of (**a, b**) demonstrated that β-III tubulin-positive fibers consisted of ND-GFP-expressing cells. White bar: 500 mm. (**d**) β-III tubulin (white) immunofluorescence staining of the hair follicle sensory nerve histocultured for 7 and 28 days. The β-III tubulin-positive fibers grew out from the whisker sensory nerve stump of the hair follicle histocultured for 7 days. By 28 days of culture, the β-III tubulin-positive fibers extended widely and radially around the hair follicle. White bar: 500 mm. (**e**) Tips of the β-III tubulin (white)-positive fibers had phalloidin (red)-positive F-actin. The presence of F-actin indicates that the tips are axon growth cones. These images are high-magnification images of the area inside the red box in (**d**). White arrows: Tips of the β-III tubulin-positive fibers. White bar: 10 mm [12]

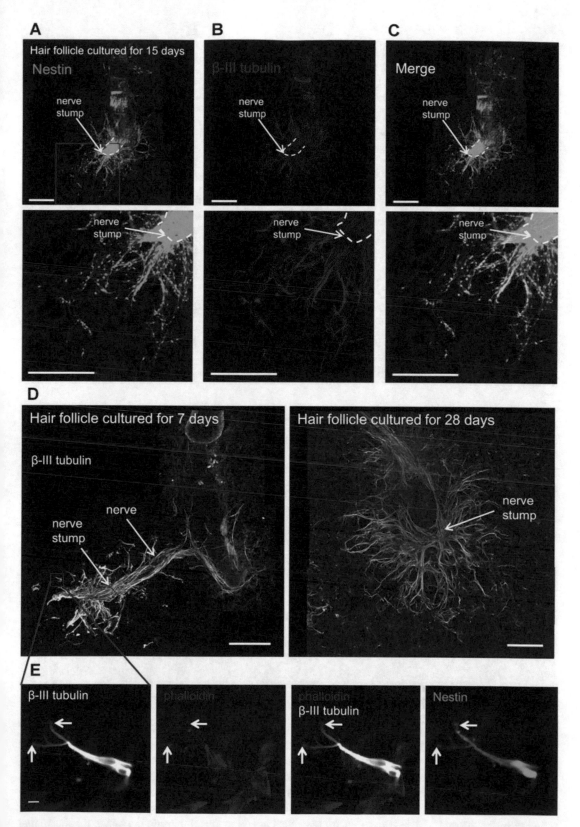

Fig. 9 Fibers of cord-like structures extending from the nerve stump expressed β-III tubulin and contained tips expressing phalloidin-positive F-actin. (**a**) A vibrissa hair follicle with its sensory nerve from an ND-GFP mouse was histocultured for 15 days on Gelfoam®. The nerve stump became enriched with HAP stem cells and many cord-like structures extended from the nerve stump. White bar: 500 mm. (**b**) Immunofluorescence staining

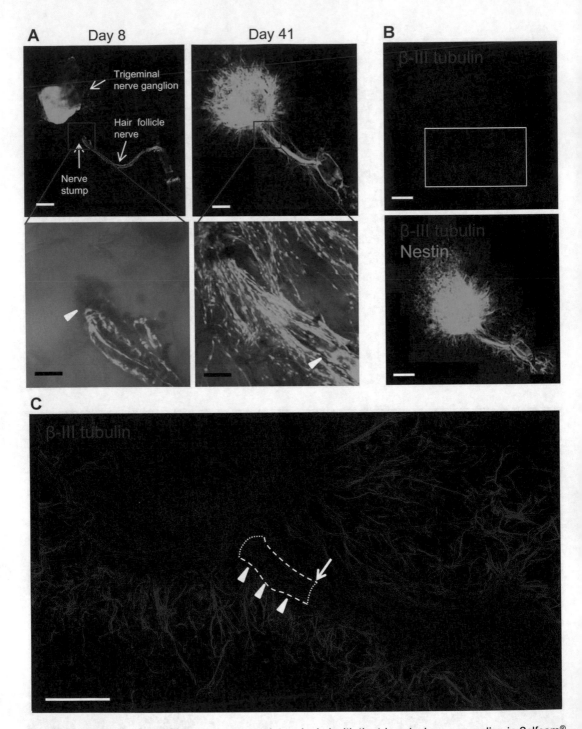

Fig. 10 The extending hair follicle sensory nerve intermingled with the trigeminal nerve ganglion in Gelfoam®histoculture. (**a**) A whisker sensory nerve of a vibrissa hair follicle was placed on Gelfoam® next to a trigeminal nerve ganglion, both from an ND-GFP transgenic mouse. By day 41, the whisker sensory nerve and the trigeminal nerve ganglion intermingled. White arrowheads: Whisker sensory nerve stump. White bar: 500 mm. Black bar: 100 mm. (**b**) Immunofluorescence staining of β-III tubulin (red) demonstrated that many β-III tubulin-positive fibers extended from both sides of the hair follicle and the trigeminal nerve ganglion. The fibers consisted of HAP stem cells. White bar: 1 mm. (**c**) Magnified image of the area inside the box in (**b**) shows that many β-III tubulin-positive fibers extended widely and radially from the trigeminal nerve ganglion and the hair follicle. There was a thick bundle of fibers linking the ganglion and the nerve (white arrowheads). White arrow: Whisker sensory nerve stump. White bar: 500 mm [12]

4.9 Nestin-Expressing Cells Participate in Nerve Growth in 3D Gelfoam® Culture

The sciatic nerve became enriched with HAP stem cells in Gelfoam® histoculture. HAP stem cells proliferated, forming fibers extending into the Gelfoam® (Fig. 11). Higher magnification images of the growing fibers demonstrated the presence of both spindle and spherical cells. The spindle cells highly expressed ND-GFP (Fig. 11). The fibers continued to extend at day 28 in Gelfoam® culture [14].

The HAP stem cells extending the nerve in Gelfoam® co-expressed p75^{NTR}, TrkB, and β-III tubulin (Fig. 12). High-magnification images show that spindle cells expressed ND-GFP, and co-expressed p75^{NTR}, TrkB, and β-III tubulin (Fig. 12). Most of the many spherical cells expressed GFAP but not nestin (Fig. 12) [14].

4.10 Fibers Extending the Sciatic Nerve in 3D Gelfoam® Histoculture Express β-III Tubulin, and Their Tips Express Phalloid in Positive F-Actin

β-III tubulin-positive fibers extended widely and radially around the nerve. The fibers consisted of ND-GFP-expressing cells (Fig. 13). The tips of the β-III tubulin-positive fibers expressed phalloidin-positive F-actin. These results suggested that β-III tubulin-positive fibers are axons growing from the sciatic nerve (Fig. 13) [14].

4.11 Inter-Nerve Co-mingling in 3D Gelfoam® Histoculture

The sciatic nerve obtained from ND-GFP transgenic mice was co-cultured on Gelfoam® with a sciatic nerve from RFP transgenic mice (Fig. 14). The ND-GFP-expressing fibers of the sciatic nerve inter-mingled with the fibers of the co-cultured RFP-expressing sciatic nerve by day 14 of histoculture (Fig. 14). The ND-GFP-expressing cells proliferated, forming radial fibers which extended the nerve. The fibers consisted of ND-GFP-expressing spindle cells and ND-GFP-negative spherical cells (Fig. 14). ND-GFP-expressing fibers invaded deeply into the RFP-expressing sciatic nerve (Fig. 14) [14].

4.12 Interaction of the Extending Sciatic Nerve and the Dorsal Root Ganglion in 3D Gelfoam® Histoculture

The sciatic nerve was co-cultured on Gelfoam® with a dorsal root ganglion, both isolated from an ND-GFP transgenic mouse (Fig. 15). Many GFP-expressing fibers were imaged extending from both the sciatic nerve and the co-cultured dorsal root ganglion at day 38 of histoculture (Fig. 15). Immunofluorescence staining also demonstrated that many β-III tubulin-positive fibers extended from both the dorsal root ganglion and the sciatic nerve. The fibers consisted of ND-GFP-expressing cells (Fig. 15). β-III tubulin-positive fibers extending from both the sciatic nerve and the dorsal root ganglion intermingled each other (Fig. 15) [14] (*see* **Notes 2, 3**).

4.13 ND-GFP-Expressing Cells Are Located in Peripheral Dorsal Root Nerves But Not in the Spinal Cord in 3D Gelfoam® Histoculture

The spinal cord with posterior roots, excised from an ND-GFP transgenic mouse, was put in Gelfoam® histoculture. On day 0, some ND-GFP-expressing cells were seen in the origin of posterior roots (Fig. 16). At day 7, the posterior roots were enriched with ND-GFP-expressing cells but were not observed in the cortex of the spinal cord (Fig. 16) [14].

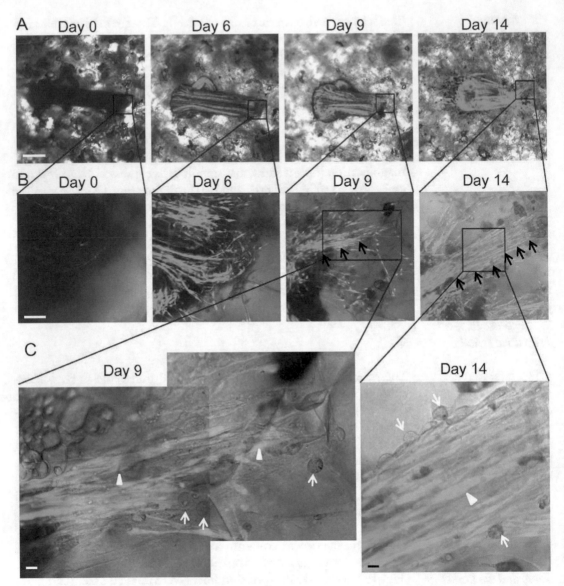

Fig. 11 ND-GFP-expressing cells proliferated in the sciatic nerve and formed fibers in 3D Gelfoam® histoculture. (**a**) Time-course imaging of HAP stem cell trafficking in the histocultured sciatic nerve from an ND-GFP
mouse. The nerve became enriched with HAP stem cells by day 14. At day 6, the HAP stem cells proliferated
in extending fibers and migrated into the Gelfoam®. At day 14, many fibers consisting of HAP stem cells
extended widely and radially around the nerve. Bar: 500 µm. (**b**) Magnified images of the growing nerve. At day
6, the HAP stem cells proliferated as fibers and migrated into the Gelfoam®. Bar: 100 µm. (**c**) High-magnification
images of the area inside the boxes in Fig. 3b. At day 9, the HAP stem cells grew in fibers (white arrowheads).
Some spherical cells proliferated in the growing fibers. They did not express nestin (white arrows). At day 14,
the spindle cells expressed ND-GFP (white arrowhead). In addition, there were spherical cells without nestin
expression (white arrow). Bar: 10 µm [14]

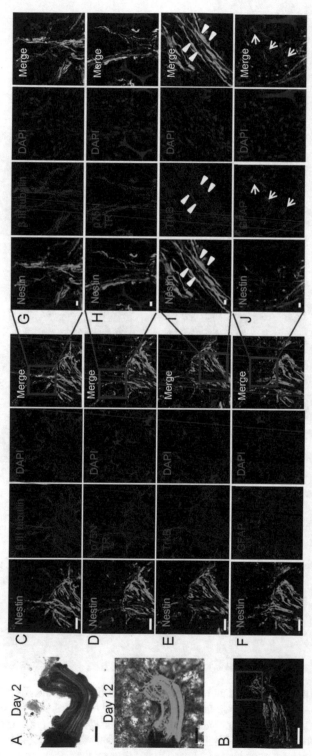

Fig. 12 Nestin-positive spindle cells expressed neuronal markers and nestin-negative spherical cells expressed glial marker in growing nerve fibers in 3D Gelfoam® histoculture. (**a**) A sciatic nerve was cultured on Gelfoam®. At day 12, the sciatic nerve was enriched with HAP stem cells and many fibers consisting of HAP stem cells extended from the growing nerve. Bar: 500 μm. (**b**) One section from **A** on day 12. Bar: 500 μm. (**c**–**f**) Magnified images from the box in (**b**). HAP stem cells in the sciatic nerve co-expressed β-III tubulin, p75^{NTR}, and TrkB. Bar: 100 μm. (**g**, **h**) Magnified images from the boxes in (**c**, **d**). There were fibers extending from the nerve. These fibers consisted of HAP stem spindle cells. The spindle cells expressing ND-GFP co-expressed β-III tubulin and p75^{NTR}. Bar: 10 μm. (**i**) High-magnification images of the area inside the box in (**e**) show that the spindle cell expressed ND-GFP in the nerve and co-expressed TrkB. Bar: 10 μm. (**j**) Magnified images from the area inside the box in (**f**). Spherical cells were observed in the growing nerve. Most of the spherical cells expressed GFAP, but not ND-GFP. Bar: 10 μm [14]

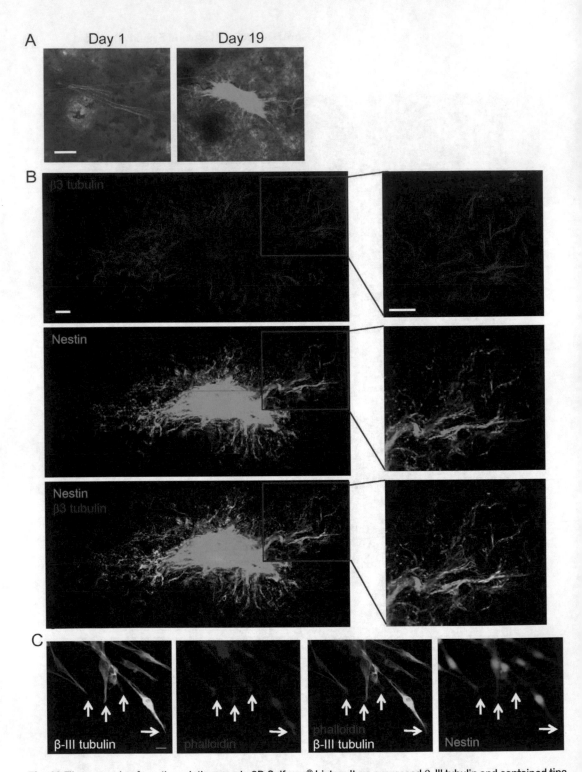

Fig. 13 Fibers growing from the sciatic nerve in 3D Gelfoam® histoculture expressed β-III tubulin and contained tips expressing phalloidin-positive F-actin. (**a**) A sciatic nerve from an ND-GFP mouse was cultured for 19 days on Gelfoam®. The nerve became enriched with HAP stem cells and many fibers extended from the growing nerve. Bar: 500 μm. (**b**) Immunofluorescence-stained images of β-III tubulin (red) demonstrated that many β-III tubulin-positive fibers extended from the nerve stump and extended widely and radially around the nerve. Merged images of Fig. 13 demonstrated that β-III tubulin-positive fibers consisted of ND-GFP-expressing cells. Bar: 500 μm. (**c**) Tips of the β-III tubulin (white)-positive fibers had phalloidin (red)-positive F-actin. The presence of F-actin indicates that tips are growth cones of axons. White arrows: Tips of the β-III tubulin-positive fiber. Bar: 10 μm [14]

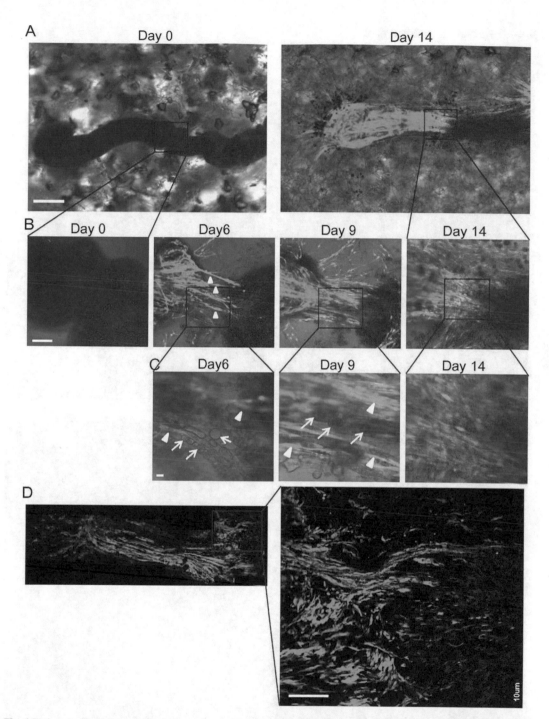

Fig. 14 Intermingling of growing sciatic nerves in 3D Gelfoam® histoculture. (**a**) A sciatic nerve from an ND-GFP trans-genic mouse was placed on Gelfoam® next to the sciatic nerve from an RFP transgenic mouse. At day 14, the sciatic nerve from the ND-GFP mouse was enriched with ND-GFP-expressing stem cells and intermingled with the RFP-expressing sciatic nerve. Bar: 500 μm. (**b**) Magnified images of the area inside the box in **A** show that the ND-GFP-expressing stem cells proliferated in fibers growing from the nerve extending toward the other sciatic nerve. At day 9, the thickest fibers appeared between both sciatic nerves. At day 14, the two nerves intermingled with each other. Bar: 100 μm. (**c**) Magnified images of the area inside the box in **B** show that the fibers consisted of ND-GFP-expressing stem spindle cells (white arrowheads) and ND-GFP-negative spherical cells (white arrows). The spherical cells formed a line between both sciatic nerves and ND-GFP-expressing stem spindle cells extended among the lines. Bar: 10 μm. (**d**) A section of the intermingling two nerves. High-magnification images show that ND-GFP-expressing stem fibers growing from the sciatic nerve from the ND-GFP mouse invaded deeply into the RFP-expressing sciatic nerve. Bar: 100 μm [14]

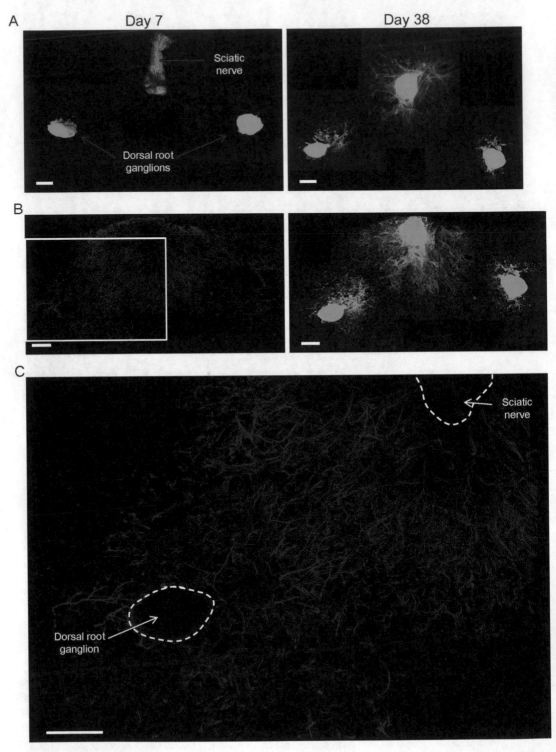

Fig. 15 The growing sciatic nerve intermingled with the dorsal root ganglion in 3D Gelfoam® histoculture. (**a**) A sciatic nerve was placed in Gelfoam® histoculture next to a dorsal root ganglion, both from an ND-GFP mouse. At day 38, many ND-GFP-expressing fibers were seen extending from both the sciatic nerve and co-cultured dorsal root ganglion. Bar: 500 μm. (**b**) Immunofluorescence staining of β-III tubulin (red) demonstrated that many β-III tubulin-positive fibers extended from both the sciatic nerve and the dorsal root ganglions.

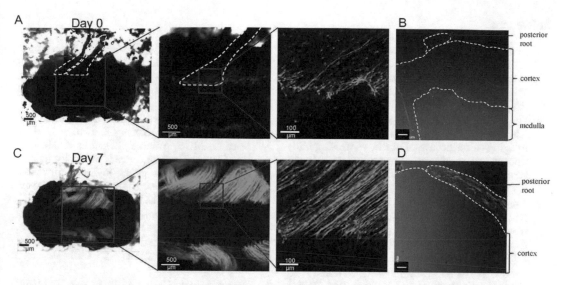

Fig. 16 ND-GFP-expressing cells are located in dorsal peripheral nerve roots but not in the spinal cord in Gelfoam® histoculture. (**a**) A spinal cord with posterior roots removed from an ND-GFP transgenic mouse was put in Gelfoam® histoculture. On day 0, some ND-GFP-expressing cells were observed in the origin of posterior root (enclosed by white dashed lines). (**b**) Cross section of the spinal cord just after removal shows that there were a few ND-GFP-expressing cells in the posterior root. White bar: 50 μm. (**c**) At day 7, posterior roots were enriched with ND-GFP-expressing cells. (**d**) Cross section of the spinal cord cultured for 7 days shows that ND-GFP-expressing cells proliferated in the posterior root but not in the cortex of spinal cord. White bar: 50 μm [14]

5 Notes

1. It is noteworthy that in the trigeminal nerve in Gelfoam® histoculture and in the regenerating trigeminal nerve in vivo, nestin-expressing cells were very prominent [12].

2. The results suggest that Gelfoam® histoculture is a physiologic system to support nerve growth and intermingling and should have broad application and enable future studies of the elongation and joining of many types of peripheral nerves to improve this process for clinical application [14].

3. The results reviewed in the present chapter indicate that Gelfoam® histoculture supports an interacting nerve organoid system. Future experiments should investigate nerve conductions in this system. For a further discussion about organoids and controversies, please see the Afterward of the present volume.

Fig. 15 (continued) The fibers consisted of ND-GFP-expressing stem cells. Bar: 500 μm. (**c**) Magnified image of the area inside the box in (**b**) shows that many β-III tubulin-positive fibers extended widely and radially both from the sciatic nerve and the dorsal root ganglion. β-III tubulin-positive fibers from both nerves intermingled with each other. Bar: 500 μm [14]

References

1. Li L, Mignone J, Yang M, Matic M, Penman S, Enikolopov G, Hoffman RM (2003) Nestin expression in hair follicle sheath progenitor cells. Proc Natl Acad Sci U S A 100: 9958–9961

2. Duong J, Mii S, Uchugonova A, Liu F, Moossa AR, Hoffman RM (2012) Real-time confocal imaging of trafficking of nestin-expressing multipotent stem cells in mouse whiskers in long-term 3-D histoculture. In Vitro Cell Dev Biol Anim 48:301–305

3. Amoh Y, Li L, Katsuoka K, Penman S, Hoffman RM (2005) Multipotent nestin-positive, keratin-negative hair-follicle-bulge stem cells can form neurons. Proc Natl Acad Sci U S A 102:5530–5534

4. Amoh Y, Li L, Campillo R, Kawahara K, Katsuoka K, Penman S, Hoffman RM (2005) Implanted hair follicle stem cells form Schwann cells that support repair of severed peripheral nerves. Proc Natl Acad Sci U S A 102:17734–17738

5. Amoh Y, Li L, Katsuoka K, Hoffman RM (2008) Multipotent hair follicle stem cells promote repair of spinal cord injury and recovery of walking function. Cell Cycle 7:1865–1869

6. Yu H, Fang D, Kumar SM, Li L, Nguyen TK, Acs G, Herlyn M, Xu X (2006) Isolation of a novel population of multipotent adult stem cells from human hair follicles. Am J Pathol 168:1879–1888

7. Yu H, Kumar SM, Kossenkov AV, Showe L, Xu XW (2010) Stem cells with neural crest characteristics derived from the bulge region of cultured human hair follicles. J Invest Dermatol 130:1227–1236

8. Amoh Y, Kanoh M, Niiyama S, Kawahara K, Satoh Y, Katsuoka K, Hoffman RM (2009) Human and mouse hair follicles contain both multipotent and monopotent stem cells. Cell Cycle 8:176–177

9. Liu F, Uchugonova A, Kimura H, Zhang C, Zhao M, Zhang L, Koenig K, Duong J, Aki R, Saito N, Mii S, Amoh Y, Katsuoka K, Hoffman RM (2011) The bulge area is the major hair follicle source of nestin-expressing pluripotent stem cells which can repair the spinal cord compared to the dermal papilla. Cell Cycle 10:830–839

10. Uchugonova A, Duong J, Zhang N, König K, Hoffman RM (2011) The bulge area is the origin of nestin-expressing pluripotent stem cells of the hair follicle. J Cell Biochem 112:2046–2050

11. Philpott MP, Green MR, Kealey T (1990) Human hair growth in vitro. J Cell Sci 97(pt 3):463–471

12. Mii S, Duong J, Tome Y, Uchugonova A, Liu F, Amoh Y, Saito N, Katsuoka K, Hoffman RM (2013) The role of hair follicle nestin-expressing stem cells during whisker sensory-nerve growth in long-term 3D culture. J Cell Biochem 114:1674–1684

13. Hoffman RM (2010) Histocultures and their uses. In: Encyclopedia of life sciences. Wiley, Chichester. https://doi.org/10.1002/9780470 015902.a0002573.pub2

14. Mii S, Uehara F, Yano S, Tran B, Miwa S, Hiroshima Y, Amoh Y, Katsuoka K, Hoffman RM (2013) Nestin-expressing stem cells promote nerve growth in long-term 3-dimensional Gelfoam®-supported histoculture. PLoS One 8:e67153

Histoculture and Infection with HIV of Functional Human Lymphoid Tissue on Gelfoam®

Andrea Introini, Wendy Fitzgerald, Christophe Vanpouille, and Leonid Margolis

Abstract

Gelfoam® histoculture provides a valuable tool for experimental studies of normal and pathological tissue physiology. It allows us to understand cell-cell interactions by mirroring their original spatial relationship within body tissues. Gelfoam® histoculture can be employed to model host-pathogen interactions mimicking in vivo conditions in vitro. In the present chapter, we describe a protocol to process and infect lymphoid tissue explants with HIV and maintain them in Gelfoam® histoculture at the liquid-air interface. The Gelfoam® histocultures with human immunodeficiency virus (HIV) type 1-infected tissues have been used to further understand the biology of early HIV-1 pathogenesis, as well as a novel ex vivo platform to test the efficacy and toxicity of antiviral drugs.

Key words Lymphoid tissue, Tonsils, Histoculture, Gelfoam®, HIV-1 infection, Antibody production

1 Introduction

Many accomplishments in cellular and molecular biology were achieved in monolayer culture. On the other hand, such monotypic two-dimensional cell cultures do not account for the spatial and functional communication between the large variety of cell types that comprise tissues and organs. This aspect is of crucial importance for experimental models of various diseases, as the interference with homeostatic cell-cell interactions is the leading factor of many pathologies. For example, the critical events of HIV infection take place in lymphoid tissue where this virus infects a small fraction of lymphocytes, but infection leads to the death of uninfected cells ("bystander death") [1] and deterioration of the entire structure of lymphoid tissue [2].

Paradoxically, tissue explant cultures (histocultures) preceded conventional cell cultures. Ross Harrison, an American biologist, performed the first successful work on tissue culture in the early

Robert M. Hoffman (ed.), *3D Sponge-Matrix Histoculture: Methods and Protocols*, Methods in Molecular Biology, vol. 1760, https://doi.org/10.1007/978-1-4939-7745-1_17, © Springer Science+Business Media, LLC, part of Springer Nature 2018

1900s [3]. Shortly after, Alexis Carrel successfully cultured chick embryo heart fragments for 3 months, thereby laying down the foundations of three-dimensional tissue culture [4]. Later, researchers switched to single cells cultures and it took about 50 years before the next improvement of explant culture technique took place: Joseph Leighton introduced the idea of a sponge matrix as a substrate and then Hoffman et al. in the 1980s developed the three-dimensional histoculture method for anticancer drug studies [5].

In the middle of the 1990s, our laboratory pioneered the technique of human lymphoid Gelfoam® histoculture, mostly for the purpose of studying the pathogenesis of HIV [6], where lymphoid tissue is a major site of virus replication.

The system of lymphoid Gelfoam® histoculture offers major advantages over monolayer cultures as explants retain tissue cytoarchitecture and many important functional aspects of cell-cell interactions in vivo. For example, upon challenge with (recall) antigens, such as diphtheria toxoid or tetanus toxoid, Gelfoam® cultured lymphoid tissue responds with a vigorous production of specific antibodies [7]. Like any other ex vivo model, human lymphoid tissue culture has its own limitations: these include donor variability, problems with tissue polarization, tissue survival that is limited to ~3 weeks, and difficulty in monitoring cells beyond the depth of confocal microscopy.

Nevertheless, human lymphoid Gelfoam® histoculture remains a model of choice to study homeostatic and pathogenic immunological processes in humans, including host-pathogen interactions (reviewed in [8]). In particular, Gelfoam® histoculture of human lymphoid tissue allows HIV-1 infection and transmission to be studied under controlled laboratory conditions. Upon inoculation ex vivo, lymphoid Gelfoam® histoculture supports productive HIV infection without exogenous cell activation, and retains the pattern of expression of key cell-surface molecules relevant to HIV infection [9, 10]. The preservation of functional structures (e.g., follicles and germinal centers) within Gelfoam® histoculture offers a unique opportunity to integrate spatial and functional aspects of the infection, providing a potential insight into the mechanisms regulating viral persistence. Lymphoid Gelfoam® histoculture was also effectively used for preclinical evaluations of potential antivirals [11–13].

In the present chapter, we describe a detailed protocol for the Gelfoam® histoculture of human lymphoid tissue obtained from tonsillar specimens, from tissue dissection to histoculture on Gelfoam®, as well as infection of Gelfoam® histoculture with HIV-1. This technique, unlike culture of tissues fully immersed in medium, provides more oxygen while delivering nutrients from the bottom of the Gelfoam® through sponge capillaries. These properties together with the natural collagen structure of the sponge significantly delay necrosis of the center of the explants, that is typical for immersed tissues.

2   Materials

2.1   Preparation of Gelfoam®

1. Sterile collagen sponges. In the protocol we refer to Gelfoam® Absorbable Collagen Sponge USP, 12–7 mm format (Pfizer, NDC: 0009-0315-08) (Gelfoam®) as an example (see **Note 1**).

2. Sterile Petri dishes, 100 mm × 20 mm or wider.

3. Culture medium (CM): RPMI1640 with L-glutamine and phenol red, 100 μM Modified Eagle's medium (MEM)-nonessential amino acids (FisherThermoScientific), 1 mM sodium-pyruvate, 50 μg/mL gentamycin sulfate, 2.5 μg/mL amphotericin B (FisherThermoScientific), 15% fetal bovine serum (FBS). We recommend testing different batches of FBS before purchasing a large stock (see **Note 2**).

4. Sterile metal forceps or tweezers.

5. Sterile metal scissors.

6. Sterile flat weighing metal spatula.

7. Six-well culture plates.

8. Water-jacketed CO_2 incubator, set at 37 °C, 5% CO_2, ≥90% humidity.

9. Water bath.

2.2   Processing of Human Tonsillar Tissue

1. Tissue transportation medium: sterile phosphate buffer saline (PBS) or other solution at physiological pH.

2. Sterile transportation container (see **Note 3**).

3. Specimens of human tonsils obtained from routine tonsillectomy or tonsillotomies (see **Note 4**).

4. Culture medium (CM) (please see above).

5. Timentin solution (100×): add 100 mL of sterile water cell culture grade to a 3.1 g vial of Timentin® (GlaxoSmithKline, NDC: 0029-6571-26) (see **Note 5**). Aliquot and store at −20 °C.

6. Sterile Petri dish, 100 mm × 20 mm.

7. 70% ethanol solution.

8. Disinfectant solution for biological waste disposal (see **Note 6**).

9. Sterile forceps or tweezers.

10. Sterile scalpels and blades (see **Note 7**).

11. Cell-free HIV-1 viral preparation(s) (see **Note 8**).

3   Methods

All experiments are independently carried out using explants from the same tissue donor to set up experimental conditions (i.e., donor-matched). The simplest experimental design comprises a minimum

Table 1
Conditions for Gelfoam® histoculture of lympoid tissue

Number of collagen Gelfoam® histoculture pieces per 12 × 7 mm sponge	Volume of medium per well of six-well plate (ml)	Number of tissue blocks per Gelfoam® piece (corresponding to one well)	Number of replicates (wells) per experimental condition
4	3	9	Minimum: 2

of two conditions: treated vs. untreated control (e.g., HIV-infected vs. uninfected Gelfoam® histoculture). We recommend including at least two replicates for each experimental condition. Calculate the number of tissue explants and Gelfoam® sponges required by each experiment in advance in order to minimize the waste of reagents while processing tissue specimens (Table 1).

Although it is not strictly necessary to prepare Gelfoam® sponges before tissue dissection, we recommend following the order outlined here, especially when handling large amount of tissue for many experiments.

3.1 Preparation of Gelfoam®

1. Supplement CM with timentin solution (CMT) before use. Prepare enough CMT medium to dissect tonsillar tissue and for Gelfoam® histoculture. Unused thawed timentin aliquots should be discarded. Unused medium containing timentin can be stored at 4 °C for 2–4 weeks but it must be re-supplemented with timentin before use.

2. Fill a 100 mm × 20 mm Petri dish with about 20–30 ml of CMT. This setting is optimal to prepare 3 Gelfoam® 12–7 mm sponges at the time. If larger amounts of sponges are required, the use of a wider Petri dish and larger CMT volume may be more convenient and speed up the preparation.

3. Transfer the Gelfoam® sponges into the Petri dish using sterile forceps and press the sponges against the bottom of the Petri dish for about 2 min using a sterile spatula (*see* **Note 9**).

4. Use sterile scissors to cut the rehydrated Gelfoam® sponges into four pieces of equal size.

5. Transfer 1 Gelfoam® piece using forceps into each well of a six-well culture plate.

6. Add 3 ml of CMT into each well.

7. Place the plates in the incubator until tissue explants are ready to be loaded on the Gelfoam®.

3.2 Dissection of Human Tonsillar Tissue

1. Allow CM enough time to reach room temperature (RT) or put it in a water bath prewarmed at 22–37 °C.

2. Supplement CM with timentin solution (CMT) before use.

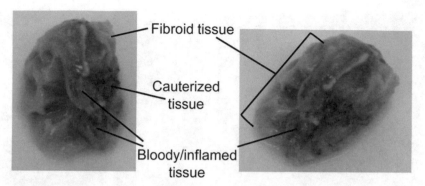

Fibroid tissue

Cauterized
tissue

Bloody/inflamed
tissue

Fig. 1 Human tonsil histoculture preparation. Tonsils are obtained from tonsillectomy or tonsillotomy. Areas that should be removed are indicated before proceeding with tissue dissection: residual fibroid tissue from the capsule surrounding the tonsil (white), cauterized tissue (brown-black), and bloody/inflamed tissue (red). Also, parts with brown-greenish color that are indicative of tissue necrosis should be removed

3. Pour 70% ethanol solution into a clean container to soak and clean forceps and scalpel whenever needed during the dissection procedure. We recommend cleaning tools and changing scalpel blade in-between handling specimens from different donors.

4. Transfer tonsils from the transportation medium into a 100 mm Petri dish containing CMT using forceps. Specimens with widespread areas of cauterized, bloody, or necrotic tissue should be discarded (*see* **Note 10**, Fig. 1).

5. The lid of the petri dish can be used as a cutting surface to dissect tissue. Holding the tissue gently with forceps, cut each tonsil into several big pieces using sterile scalpel and blade. Remove cauterized, bloody, and fibroid tissue, and any parts containing tonsilloliths (calcified material) and/or with green-brownish color. While working on one piece of tissue, it is important to keep the rest of the tissue submerged in medium to avoid tissue desiccation (*see* **Note 11**).

6. Cut a tissue piece into slices of about 2 mm in thickness. Remove any unwanted part as in the previous step. Cut the tissue slices into 2 mm-thick strips. Cut the tissue strips into 2 mm-thick blocks. This should result in tissue blocks of roughly 8 cubic mm, also referred to as explants (Fig. 2).

7. Transfer the explants into a new 100 mm Petri dish containing CMT to avoid desiccation while proceeding with tissue dissection.

8. Swirl the plate to randomize tissue explant distribution.

9. Place nine tissue explants on top of each Gelfoam® sponge in a six-well plate using forceps allowing spacing between explants (Fig. 2).

Fig. 2 Dissection and culture of human tonsillar tissue. Pieces of tonsils cleared from unwanted parts (e.g., fibroid, cauterized, bloody, necrotic as indicated in Fig. 1) were dissected into strips (upper left) and next into blocks of approximately 8 cubic mm (upper right). Tonsillar tissue explants on top of Gelfoam® sponges (nine explants/sponge) in a six-well plate containing culture medium (bottom)

10. Place the plates in the incubator (*see* **Note 12**).

11. Dispose of all biological waste in an apposite container containing disinfectant solution.

3.3 Infection of Tissue Explants

If infection is not required, go to Subheading 3.4. Typically, explants are cultured overnight and inoculation of cultures with HIV-1 is performed the next day (*see* **Note 12**).

1. Put CM in a water bath prewarmed at 37 °C.

2. Supplement CM with timentin solution (CMT) before use. Discard unused thawed timentin aliquots. Store unused medium containing timentinat 4 °C for 2–4 weeks but it must be re-supplemented with timentin before use.

3. Aspirate the medium in the six-well plate with a pipette and discard it in a container with disinfectant solution. Tilt the plate and gently push the collagen sponges to the upper part of the well to allow the medium to gather at the bottom, aspirate and discard it.

4. Add 3 ml of CMT to each well.

5. Thaw the HIV-1 viral preparation in a water bath prewarmed at 37 °C. If necessary, dilute viral stock to an appropriate concentration. It is preferable to have a dilution such that the desired inoculum is contained in a 5–8 µl volume. It is critical for an efficient infection to allow the minimum time required between thawing HIV-1 viral preparation and infecting tissue explants.

6. Pipette the inoculum directly on top of each of the nine tissue blocks of Gelfoam® (*see* **Note 13**).

7. For an accurate delivery of the inoculum, use a reverse pipetting technique and change the pipet tip for every well. Return the plate to the incubator as soon as infection is completed.

3.4 Gelfoam® Culture of Tissue Explants

Change CM of the explant culture every 3 days starting from tissue dissection (or infection). Harvest Gelfoam® histoculture supernatants and/or explants at regular intervals throughout culture time to monitor a number of parameters (*see* **Note 14**).

1. Put CM in a water bath prewarmed at 37 °C. It is not required to supplement CM with timentin.

2. Aspirate the medium in the six-well plate with a pipette and discard it in a container with disinfectant solution. Tilt the plate and gently push the collagen sponges to the upper part of the well to allow the medium to gather at the bottom, aspirate and discard it.

3. Add 3 ml of CM to each well.

4. Put back on Gelfoam® any blocks of tissue using forceps that may have fallen off from it.

5. Return the plate to the incubator.

6. Repeat **steps 1–5** every 3 days until day 12–15 post-dissection or -infection (*see* **Note 15**).

7. Dispose of the plates with any residual medium, tissue explants, and collagen sponges.

4 Notes

1. Gelfoam® sponges are hemostatic devices prepared from porcine skin, which are meant to be completely absorbed by the human body after a few weeks. Similar products are available

from a number of providers. We have chosen Gelfoam® due to its slow degradation rate in culture that better suits the purpose of maintaining tissue explants at the liquid-air interface for 2 weeks, compared to other collagen sponges that dissolve rapidly. Spongostan™ Standard, Absorbable Haemostatic Collagen Sponge 7 x 5 x 1 cm (Ferrosan Medical Devices, ref. MS0002), is a valid, perhaps almost identical, alternative that we tested and used in our experiments. In general, we recommend testing different batches of the same product in order to assess its performance in explant culture.

2. We advise testing several lots of FBS for culture optimization and use the same lot of FBS for an entire study. We routinely test FBS on tissue explants from 10 donors and select the lot that gives the highest HIV-1 replication. Of note, FBS can also affect the ability of tissue explants to secrete cytokines.

3. Pour enough liquid into the container so that the tonsils may stay submerged in it after surgery and during transportation to avoid tissue desiccation. We recommend leaving the specimens in transportation medium at room temperature.

4. The protocol for collection of human tissues requires ethical approval by local competent authorities as well as informed consent from the patient. Obtaining personal and medical data (e.g., reason for surgery, current drug use, history of infections) from tissue donors may provide important information to interpret experimental results, although it poses important concerns regarding informed consent and privacy that must be addressed in the protocol for tissue collection.

5. Timentin® is a mix of the antibiotics ticarcillin and clavulanate that is commercially available as individual reagents. These antibiotics efficiently prevent the growth of bacteria that, in our laboratory, often contaminate tissue samples. Each laboratory should determine which antibiotics work best for their purposes. For example, penicillin and streptomycin can be used instead of Timentin, although they have different properties. For example, Timentin displays low stability at room temperature or 37 °C (about 24 h); therefore, once added to culture medium, it remains active only for the first day of culture.

6. The entire procedure should be carried out in a biological safety cabinet. All human specimens, even those from "healthy" individuals, may harbor infectious agents. The operator should receive training to work with blood-borne pathogens to safely handle human tonsillar tissue specimens, even if experiments do not involve the use of exogenous infectious agents (e.g., HIV). A risk assessment of the entire procedure should be carried out by trained staff at the laboratory to ensure the safety of the operator and coworkers.

7. Handling sharp tools to dissect human tissue specimens puts the operator at more risk of accidental injury and contamination. The operator should consider wearing metal mesh cut-resistant gloves as additional protection.

8. For most of our experiments we have used the following viral preparations: HIV-1$_{BaL}$ and HIV-1$_{LAI.04}$ obtained from the clarified culture medium of peripheral blood mononuclear cell cultures inoculated with either HIV-1$_{BaL}$ or HIV-1$_{LAI.04}$, originally received from the NIH AIDS Reagent Program. A Biosafety Level 3 facility, or dedicated BSL2 laboratories with BSL3 practices, and adequate training are required to handle infectious agents.

9. Gelfoam® sponges are extremely brittle when hydrated. The hydration process should be done carefully especially when pushing down on the foam with the spatula to force the air out. When sponges are first added to medium, wet sponges on both sides then allow a few minutes for the sponges to begin soaking up medium before pushing down with any force. Fully hydrated sponges should be as free of air as possible: the presence of air will block the capillaries through which culture medium nutrients reach the tissue.

10. We recommend processing tonsils the same day of surgery. If that would not be possible, specimens may be left overnight submerged in culture medium containing antibiotics at 4 °C. Take note of any feature of tissue specimens that you suspect may affect experimental results, such as bloody, cauterized, and necrotic parts. We suggest saving an aliquot of medium in which tissue dissection is performed and/or a few tissue explants for measuring cell viability, nucleic acids extraction, and/or histological analysis.

11. During dissection, take care to handle tissue gently as cells are easily released from tonsil tissue. Keeping a small volume of medium in the petri dish while sectioning a particular piece will prevent the tissue from floating around and making dissection more difficult. But be mindful to keep the rest of the tissue covered completely in medium. Scalpel blades should be changed often to prevent unnecessary damage to tissue with dull blades.

12. We prefer performing infection after an overnight incubation to make sure that no bacterial or fungal contamination develops. In addition, cells tend to egress from tonsillar tissue blocks after dissection. This process is largely completed within the first 24 h of culture.

13. In some circumstances, it may be preferable to infect by soaking tissue blocks in a small amount of medium mixed with virus (max volume of 500 µl) in a 1.7 ml microcentrifuge tube

for 2 h at 37 °C. Tissue blocks can then either be transferred on Gelfoam®, or washed first and then transferred. Although more laborious, this way of infecting tissue explants may give better result in tissue where HIV replication is not very high, such as in cervico-vaginal tissue [14].

14. Gelfoam® histoculture offers a number of possible experimental readouts. For nucleic acid extractions or histology tissue explants can be snap-frozen, preserved in apposite solutions for RNA stabilization, or embedded in formalin. For flow cytometry we immediately process tissue blocks by digesting them with an enzymatic cocktail to obtain a single-cell suspension for staining. Even if it is not required by the experiment, we recommend saving tissue explants for storage at –80 °C once the culture is ended. Gelfoam® histoculture supernatant can be used to measure the secretion of soluble factors (e.g., cytokines) or HIV-1 replication over time. Our standard readout is to measure the concentration of the HIV-1 core protein $p24_{gag}$ in samples of culture supernatant collected every third day post-infection [15].

15. During Gelfoam® histoculture, tissue outgrows into the Gelfoam® sponges. Also individual cells may migrate out of the tissue into the sponges. One can analyze the part of tissue that is above the Gelfoam® histoculture separately by pinching it out with forceps. The cells within the Gelfoam® histoculture can be squeezed out using forceps or a syringe plunger. However, one should keep in mind that what is extracted from the Gelfoam® histoculture is not just a fraction of migrated cells but rather a mixture of such cells and tissue structures.

References

1. Biancotto A, Iglehart SJ, Vanpouille C, Condack CE, Lisco A, Ruecker E, Hirsch I, Margolis LB, Grivel JC (2008) HIV-1 induced activation of CD4+ T cells creates new targets for HIV-1 infection in human lymphoid tissue ex vivo. Blood 111:699–704

2. Lederman MM, Margolis L (2008) The lymph node in HIV pathogenesis. Semin Immunol 20:187–195

3. Harrison R (1910) The outgrowth of the nerve fiber as a mode of protoplasmic movement. J Exp Zool 9:787–846

4. Carrel A (1912) On the permanent life of tissues outside of the organism. J Exp Med 15:516–528

5. Hoffman RM (1991) Three-dimensional histoculture: origins and applications in cancer research. Cancer Cells 3:86–92

6. Glushakova S, Baibakov B, Margolis LB, Zimmerberg J (1995) Infection of human tonsil histocultures: a model for HIV pathogenesis. Nat Med 1:1320–1322

7. Glushakova S, Grivel JC, Fitzgerald W, Sylwester A, Zimmerberg J, Margolis LB (1998) Evidence for the HIV-1 phenotype switch as a causal factor in acquired immunodeficiency. Nat Med 4:346–349

8. Grivel JC, Margolis L (2009) Use of human tissue explants to study human infectious agents. Nat Protoc 4:256–269

9. Grivel JC, Margolis LB (1999) CCR5- and CXCR4-tropic HIV-1 are equally cytopathic for their T-cell targets in human lymphoid tissue. Nat Med 5:344–346

10. Grivel JC, Penn ML, Eckstein DA, Schramm B, Speck RF, Abbey NW, Herndier B, Margolis

L, Goldsmith MA (2000) Human immuno-deficiency virus type 1 coreceptor preferences determine target T-cell depletion and cellular tropism in human lymphoid tissue. J Virol 74:5347–5351

11. Lisco A, Vanpouille C, Tchesnokov EP, Grivel JC, Biancotto A, Brichacek B, Elliott J, Fromentin E, Shattock R, Anton P, Gorelick R, Balzarini J, McGuigan C, Derudas M, Gotte M, Schinazi RF, Margolis L (2008) Acyclovir is activated into a HIV-1 reverse transcriptase inhibitor in herpesvirus-infected human tissues. Cell Host Microbe 4:260–270

12. Vanpouille C, Lisco A, Introini A, Grivel JC, Munawwar A, Merbah M, Schinazi RF, Derudas M, McGuigan C, Balzarini J, Margolis L (2012) Exploiting the anti-HIV-1 activity of acyclovir: suppression of primary and drug-resistant HIV isolates and potentiation of

the activity by ribavirin. Antimicrob Agents Chemother 56:2604–2611

13. Vanpouille C, Khandazhinskaya A, Karpenko I, Zicari S, Barreto-de-Souza V, Frolova S, Margolis L, Kochetkov S (2014) A new antiviral: chimeric 3TC-AZT phosphonate efficiently inhibits HIV-1 in human tissues ex vivo. Antiviral Res 109:125–131

14. Saba E, Grivel JC, Vanpouille C, Brichacek B, Fitzgerald W, Margolis L, Lisco A (2010) HIV-1 sexual transmission: early events of HIV-1 infection of human cervico-vaginal tissue in an optimized ex vivo model. Mucosal Immunol 3:280–290

15. Biancotto A, Brichacek B, Chen SS, Fitzgerald W, Lisco A, Vanpouille C, Margolis L, Grivel JC (2009) A highly sensitive and dynamic immuno-fluorescent cytometric bead assay for the detection of HIV-1 p24. J Virol Methods 157:98–101

Chapter 18

Imaging DNA Repair After UV Irradiation Damage of Cancer Cells in Gelfoam® Histoculture

Shinji Miwa and Robert M. Hoffman

Abstract

DNA damage repair in response to UVC irradiation was imaged in cancer cells growing in Gelfoam® histoculture. UVC-induced DNA damage repair was imaged with green fluorescent protein (GFP) fused to the DNA damage response (DDR)-related binding protein 53BP1 in MiaPaCa-2 human pancreatic cancer cells. Three-dimensional Gelfoam® histocultures and confocal imaging enabled 53BP1-GFP nuclear foci to be observed within 1 h after UVC irradiation, indicating the onset of DNA damage repair response. Induction of UV-induced 53BP1-GFP focus formation was limited up to a depth of 40 μm in Gelfoam® histoculture of MiaPaCa-2 cells, indicating this was the depth limit of UVC irradiation.

Key words Cancer cells, Gelfoam®, Histoculture, UV irradiation, DNA damage repair, 53PB1-GFP, Imaging

1 Introduction

We review in this chapter imaging DNA repair using 53BP1-GFP focus formation as a marker of early response to DNA damage in cancer cells in Gelfoam® histoculture [1].

Previously, Efimova et al. [2] fused GFP to the chromatin-binding domain of the DNA damage response (DDR)-related checkpoint-adapter protein 53BP1 and observed focus formation of this fusion protein after ionizing radiation (IR). The role of these proteins in the DNA repair response after UV light is poorly understood [1].

In this chapter, we review using 53BP1-GFP focus formation as an imaging marker of early response to DNA damage of cancer cells in Gelfoam® histoculture [1].

Robert M. Hoffman (ed.), *3D Sponge-Matrix Histoculture: Methods and Protocols*, Methods in Molecular Biology, vol. 1760, https://doi.org/10.1007/978-1-4939-7745-1_18, © Springer Science+Business Media, LLC, part of Springer Nature 2018

2   Materials

2.1   Materials for the Preparation of Gelfoam®

1. Gelfoam® sources: Gelfoam sponges comprised of absorbable gelatin prepared from purified porcine skin (Pharmacia & Upjohn Co., Kalamazoo, MI).

2. Gelfoam® trimming aid: Double-edged blade.

3. Sterile petri dish, 100 mm × 20 mm or wider.

4. Sterile metal forceps or tweezers.

5. Six-well culture plates.

2.2   Cell Culture

1. MiaPaCa-2[Tet-On] Advanced cell line.

2. pLVX-Tight-Puro lentiviral vector (Clontech, Mountain View, CA).

3. High-glucose DMEM (Invitrogen, Grand Island, NY).

4. Tet system-approved fetal bovine serum (Clontech, Mountain View, CA).

5. Puromycin (Thermo Fisher Scientific, Waltham, MA).

6. G418 (Thermo Fisher Scientific, Waltham, MA).

7. Doxycycline (Sigma, St. Louis, MO).

8. 35 mm Culture dishes.

9. CO_2 incubator.

10. Centrifuge.

2.3   Confocal Laser Scanning Microscopy

1. FV1000 confocal laser scanning microscope (Olympus, Tokyo, Japan).

2.4   UV Irradiation

1. Benchtop 3UV transilluminator (UVP, LLC, Upland, CA).

3   Methods

3.1   Cell Culture and Gene Constructs

1. Fuse GFP to the human 53BP1 IRIF binding domain of MiaPaCa-2 using pLVX-Tight-Puro lentiviral vector.

2. Transduce the vector into the MiaPaCa-2 Tet-On Advanced cell line.

3. Culture in high-glucose DMEM with 10% Tet system-approved fetal bovine serum in the presence of G418 (50 μg/ml) and puromycin (10 μg/ml).

4. Induce 53BP1-GFP by treatment with 1 μg/ml doxycycline 48 h before observation of GFP expression by fluorescence imaging.

3.2 Three-Dimensional Histoculture Using Gelfoam®

1. Cut Gelfoam® $10 \times 10 \times 3$ mm^3 with a sterile razor.

2. Soak the Gelfoam® in a 35 mm culture dish filled with 3 ml DMEM medium 1 µg/ml doxycycline, with sufficient volume to cover the Gelfoam®.

3. Centrifuge MiaPaCa-2^{Tet-On} 53BP1-GFP cells.

4. Put the cell pellet of MiaPaCa-2^{Tet-On} 53BP1-GFP cells (1×10^6) on the hydrated Gelfoam®.

5. Incubate the cultures at 37 °C, 5% CO_2, 100% humidity.

6. Forty-eight hours after seeding, irradiate the cells with UVC with the Benchtop 3UV transilluminator.

7. Observe GFP focus formation with confocal microscopy.

3.3 Confocal Laser Scanning Microscopy

1. Use a FV1000 confocal laser scanning microscope for two-(X, Y) and three-dimensional (3D, X, Y, Z) high-resolution imaging of focus formation within MiaPaCa-2^{Tet-On} 53BP1-GFP cells in three-dimensional culture. Obtain images of optional sections of 10 µm, each starting from the surface of the Gelfoam®. Reconstruct three-dimensional images using FV1000 software.

3.4 Statistical Analysis

1. Express the experimental data as the mean ± SD.

2. Perform statistical analysis using the Student's t-test or one-way analysis of variance (ANOVA) test.

3. Consider the P-values less than 0.05 statistically significant [1].

4 Results

4.1 Imaging 53BP1-GFP Focus Formation in MiaPaCa-2Tet-On 53BP1-GFP Cells in Three-Dimensional Gelfoam® Histoculture After UVC Irradiation

To investigate the depth of penetration by UVC irradiation, 53BP1GFP focus formation was determined in three-dimensional Gelfoam® histoculture [3–6]. One hour after 500 J/m^2 UVC irradiation, 53BP1-GFP focus formation of the cells at each depth was imaged using the FV1000 confocal microscope (Fig. 1). The cells at 20 and 40 µm depth had increased focus formation after irradiation ($P < 0.05$) (see **Notes 1 and 2**). However, there was no significant difference in focus formation in cells at 60 and 80 µm depths and control cells (see **Note 2**). This result suggests that penetration of UVC is limited to 40 µm depth in Gelfoam® (see **Notes 3 and 4**) [1].

5 Notes

1. The results of this study demonstrate that 53BP1-GFP is a marker of UV-induced DNA damage repair that can be imaged in Gelfoam® histoculture [1].

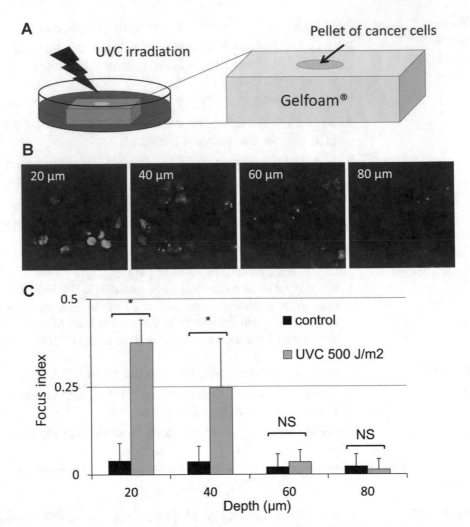

Fig. 1 Penetration of UVC in three-dimensional Gelfoam® histoculture of MiaPaCa-2[Tet-On] 53BP1-GFP cells. (**a**) MiaPaCa-2[Tet-On] 53BP1-GFP cells were centrifuged and the cell pellet was put on Gelfoam® and irradiated with UVC. One hour after UVC irradiation, 53BP1-GFP focus formation at the designated depth of Gelfoam® was imaged. (**b**) The UVC-irradiated cells at 20 and 40 μm depth showed increased focus formation compared to untreated controls, but the cells at 60 and 80 μm showed only small numbers of foci. (**c**) There were significant differences in the focus index between the UVC-treated cultures and control cultures at only 20 and 40 μm depth. The experimental data are expressed as the mean ± SD. Statistical analysis was performed using the Student's *t*-test. *$P < 0.05$, compared with control [1]

2. UVC can kill superficial cancer cells up to a depth of 40 μm [1].

3. The 40 μm depth of UVC penetration may also pertain in vivo since Gelfoam® histoculture is in vivo-like [1].

4. The results reviewed in the present chapter suggest Gelfoam® histoculture can be used to screen for DNA damaging agents.

References

1. Miwa S, Yano S, Hiroshima Y, Tome Y, Uehara F, Mii S, Efimove EV, Kimura H, Hayashi K, Tsuchiya H, Hoffman RM (2013) Imaging UVC-induced DNA damage response in models of minimal cancer. J Cell Biochem 114:2493–2499

2. Efimova EV, Mauceri HJ, Golden DW, Labay E, Bindokas VP, Darga TE, Chakraborty C, BarretoAndrade JC, Crawley C, Sutton HG, Kron SJ, Weichselbaum RR (2010) Poly (ADP-ribose) polymerase inhibitor induces accelerated senescence in irradiated breast cancer cells and tumors. Cancer Res 70:6277–6282

3. Leighton JA (1951) Sponge matrix method for tissue culture; formation of organized aggregates of cells in vitro. J Natl Cancer Inst 12:545–561

4. Freeman AE, Hoffman RM (1986) In vivo-like growth of human tumors in vitro. Proc Natl Acad Sci U S A 83:2694–2698

5. Vescio RA, Redfern CH, Nelson TJ, Ugoretz S, Stern PH, Hoffman RM (1987) In vivo-like drug responses of human tumors growing in three-dimensional gel-supported, primary culture. Proc Natl Acad Sci U S A 84:5029–5033

6. Hoffman RM (2010. Published Online) Histocultures and their use. In: Encyclopedia of life sciences. Wiley, Chichester. https://doi.org/10.1002/9780470015902.a0002573.pub2

Comparison of "Dimensionality" of Cancer Cell Culture in Gelfoam® Histoculture and Matrigel

Yasunori Tome, Fuminari Uehara, Fuminori Kanaya, and Robert M. Hoffman

Abstract

Cell and tissue culture can be performed on different substrates such as on plastic, in Matrigel™, and on Gelfoam®, a sponge matrix. Each of these substrates consists of a very different surface, ranging from hard and inflexible, a gel, and a sponge-matrix, respectively. Folkman and Moscona found that cell shape was tightly coupled to proper gene expression. The flexibility of a substrate is important for cells to maintain their optimal shape. Human osteosarcoma cells, stably expressing a fusion protein of a_v integrin, and green fluorescent protein (GFP), grew as a simple monolayer without any structure formation on the surface of a plastic dish. When the osteosarcoma cells were cultured within Matrigel, the cancer cells formed colonies but no other structures. When the cancer cells were seeded on Gelfoam®, the cells formed 3-dimensional tissue-like structures. These results indicate that Gelfoam® histoculture, unlike Matrigel™ culture, is true 3-dimensional.

Key words Cancer cells, Monolayer, Matrigel™, Gelfoam®, Green fluorescent protein (GFP), Fluorescence imaging, Rendering, Dimensionality, Aggregation, Tissue, Structures

1 Introduction

1.1 Gelfoam® Histoculture

Leighton developed sponge-gel matrix histoculture [1–15] (*see* Foreword and Chapter 1 in the present volume). Leighton made a number of important early observations on the advantages of sponge-matrix histoculture; for example, when C3HBA mouse mammary adenocarcinoma cells were grown in sponge-matrix histoculture, he found that the cells aggregated in a manner similar to that in the original tumor. Distinct structures were formed within the tumors such as lumina and stromal elements, with some of the structures similar to the original tumor [1–15] (*see* Foreword and Chapter 1 in the present volume).

In the current century, articles have appeared about "the new dimension" in cell culture [16, 17], due to the use of a mouse cancer-cell extracellular matrix preparation called "Matrigel™",

Robert M. Hoffman (ed.), *3D Sponge-Matrix Histoculture: Methods and Protocols*, Methods in Molecular Biology, vol. 1760, https://doi.org/10.1007/978-1-4939-7745-1_19, © Springer Science+Business Media, LLC, part of Springer Nature 2018

reflecting the authors' ignorance of Leighton's 3-dimensional sponge-gel histoculture and applications described in the present volume representing more than 50 years of published results.

Matrigel™ [18, 19] was first isolated from a murine tumor as a crude protein mixture that was a liquid at 4 °C and gelled at 24–37 °C. Matrigel™ comprises mainly laminin-111, collagen IV, heparin sulfate proteoglycan, various growth factors, and additional components [20]. In Matrigel™, cancer cells formed branched and invasive structures while the less malignant cells formed small aggregates. Invasive morphology correlated with the metastatic ability of the cancer cells in vivo [20].

Folkman and Moscona [21] observed that cell shape is critical for DNA synthesis and cell proliferation. It would seem important that a culture substrate should let cells acquire their natural shape. Therefore, we directly compared cancer cell growth on plastic, Matrigel™, and Gelfoam®.

We review here the use of green fluorescent protein (GFP) imaging to visualize and compare the behavior of osteosarcoma cells in monolayer culture, Matrigel™ culture, and sponge-matrix Gelfoam® culture. We observed that each substrate provided a very different result with three-dimensional tissue-like structures formed on Gelfoam® and not Matrigel™ or plastic [22].

2 Materials

2.1 Establishment of Human Osteosarcoma Cells Expressing α_v Integrin-GFP

1. Human 143B α_v integrin-GFP-expressing osteosarcoma cell line.
2. Cell culture medium: RPMI 1640 medium (Irvine Scientific, Santa Ana, CA).
3. Fetal bovine serum (FBS) (Omega Scientific, San Diego, CA).
4. 1% Penicillin/streptomycin (Sigma-Aldrich, St. Louis, MO).
5. pCMV6-AC-ITGAV-GFP vector (OriGene Technologies, Rockville, MD).
6. Lipofectamine LTX (Invitrogen, Carlsbad, CA).
7. G418 (Sigma-Aldrich, St. Louis, MO).

2.2 Monolayer Culture

1. Fibronectin-coated culture dish (35 mm) (BD Pharmingen, San Diego, CA).
2. Cell culture medium: RPMI 1640 medium (Irvine Scientific, Santa Ana, CA).
3. Fetal bovine serum (FBS) (Omega Scientific, San Diego, CA).
4. Penicillin/streptomycin (1%) (Sigma-Aldrich, St. Louis, MO).

**2.3 Matrigel™
Culture**

1. Matrigel™ (BD Biosciences).

2. Cell culture medium: RPMI 1640 medium (Irvine Scientific, Santa Ana, CA).

3. Fetal bovine serum (FBS) (Omega Scientific, San Diego, CA).

4. 1% Penicillin/streptomycin (Sigma-Aldrich, St. Louis, MO).

5. 143B α_v integrin-GFP cell line.

**2.4 Gelfoam®
Histoculture**

1. Gelfoam® Absorbable Collagen Sponge USP, 12–7 mm format (Pfizer, NDC: 0009-0315-08) (Gelfoam®) as an example.

2. Cell culture medium: RPMI 1640 medium (Irvine Scientific, Santa Ana, CA).

3. Fetal bovine serum (FBS) (Omega Scientific, San Diego, CA).

4. 1% Penicillin/streptomycin (Sigma-Aldrich, St. Louis, MO).

5. 143B α_v integrin-GFP cell line.

**2.5 Fluorescence
Imaging (See Notes 2,
3, and 4)**

1. Olympus FV1000 laser-scanning confocal microscope (Olympus, Tokyo, Japan).

2. XLUMPLFL 20× (0.95 NA) water-immersion objective (Olympus, Tokyo, Japan).

3. FV 10-ASW Fluoview software (Olympus, Tokyo, Japan).

4. ImageJ (NIH, Bethesda, MD).

3 Methods

**3.1 Establishment
of Human
Osteosarcoma Cells
Expressing α_v
Integrin-GFP**

1. Use human osteosarcoma cell line, 143B. Maintain the cells with RPMI 1640 medium containing 10% FBS and 1% penicillin/streptomycin at 37 °C in a humidified incubator with 5% CO_2.

2. Transfect cells with the α_v integrin-GFP fusion vector, pCMV6-AC-ITGAV-GFP (OriGene Technologies, Rockville, MD), using Lipofectamine LTX (Invitrogen, Carlsbad, CA) according to the manufacturer's instruction.

3. After 24-h transfection, select cells stably expressing α_v integrin-GFP with RPMI 1640 medium containing 10% FBS, 1% penicillin/streptomycin, and 800 µg/mL G418 (Sigma-Aldrich, St. Louis, MO) at 37 °C in a humidified incubator with 5% CO_2

4. Maintain 143B α_v integrin-GFP cells with RPMI 1640 medium containing 10% FBS, 1% penicillin/streptomycin, and 500 µg/mL G418 at 37 °C in a humidified incubator with 5% CO_2 (*see* **Note 1**).

A

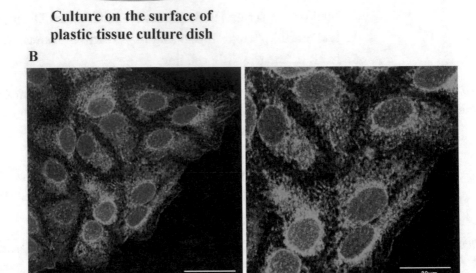

Culture on the surface of plastic tissue culture dish

B

Fig. 1 Behavior of 143B cells expressing a_v integrin-GFP in monolayer culture. (a) Schematic of culture of 143B human osteosarcoma cells on flat, plastic substrates. (b) Confocal fluorescence micrographs of 143B human osteosarcoma cells, expressing a_v integrin-GFP, in monolayers on plastic dishes [22]

3.2 Monolayer Culture

1. Seed 143B cells (1×10^4) stably expressing α_v integrin-GFP in 35 mm fibronectin-coated dishes (BD Pharmingen) for 24–48 h (*see* **Note 1**).

2. After 24–48-h seeding, observe cells using Olympus FV 1000 laser-scanning confocal microscope (Olympus, Tokyo, Japan) with an XLUMPLFL 20× (0.95 NA) water-immersion objective (Olympus, Tokyo, Japan) (*see* **Note 2**) (Fig. 1).

3.3 Matrigel™ Culture

1. Thaw Matrigel™ (BD Biosciences) on ice at room temperature gradually prior to use.

2. Coat cell culture plates (35 mm) with 600 uL Matrigel™ and incubate at 37 °C for 30 min.

3. Seed 143B-expressing α_v integrin-GFP cells (1×10^4) on the bottom of a cell culture plate coated with Matrigel™ (*see* **Note 1**).

4. Allow 143B-expressing α_v integrin-GFP cells to spread for 2–3 h.

5. Add Matrigel™ (5%) with RPMI 1640 medium to cell culture.

6. Incubate cells at 37 °C in a humidified incubator with 5% CO_2.

A

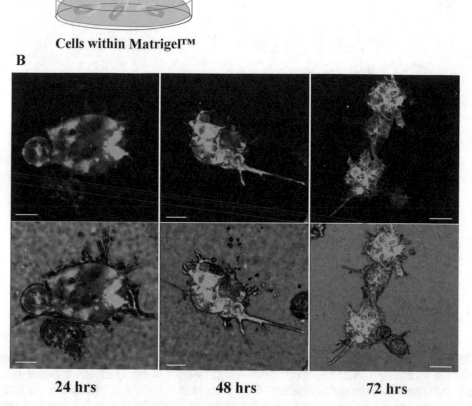

Cells within Matrigel™

B

24 hrs   48 hrs   72 hrs

Fig. 2 Behavior of 143B cells expressing a_v integrin-GFP in Matrigel™ culture. (**a**) Schematic of culture of 143B cells sandwiched between a Matrigel™-coated surface and a layer of Matrigel™. (**b**) Phase-contrast (lower row) and fluorescence (upper row) micrographs of live human osteosarcoma cells expressing a_v integrin in Matrigel™ culture. FV1000 confocal fluorescence microscopy. Bar: 20 mm (left panels). Bar: 30 mm (middle panels). Bar: 50 mm (right panels) [22]

7. Observe cells over time using an Olympus FV 1000 laser-scanning confocal microscope (Olympus, Tokyo, Japan) with an XLUMPLFL 20× (0.95 NA) water-immersion objective (*see* **Notes 2, 3,** and **4**) (Fig. 2).

3.4 Gelfoam®
Histoculture

1. Cut Gelfoam® sponges into 1 cm cubes under sterile procedure.

2. Place the Gelfoam® cubes in six-well tissue culture plates under sterile procedure.

3. Gently add RPMI 1640 medium in six-well tissue culture plates so that Gelfoam® cubes are not moved.

4. Incubate at 37 °C in a humidified incubator with 5% CO_2 in order that the Gelfoam® absorbs the medium.

5. Seed 143B-expressing α_v integrin-GFP cells (1×10^6) on top of the hydrated Gelfoam® (*see* **Note 1**).

A

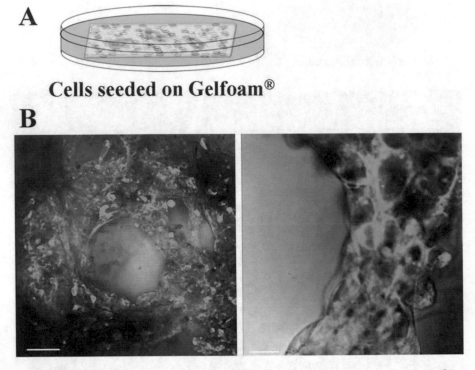

Cells seeded on Gelfoam®

B

Fig. 3 Behavior of 143B cells expressing aᵥ integrin-GFP on Gelfoam®. (**a**) Schematic of Gelfoam® histoculture of 143B cells. (**b**) Phase-contrast and fluorescence-merged images of live 143B human osteosarcoma cells expressing aᵥ integrin in Gelfoam® histoculture. FV1000 confocal fluorescence microscopy. The cancer cells formed three-dimensional complexes and gland-like structures within the Gelfoam® within 72 h (left panel). Bar: 100 mm. Diffuse expression of aᵥ integrin-GFP in the cancer cells in the Gelfoam® culture (right panel). Bar: 20 mm [22]

6. Incubate 143B-expressing α_v integrin-GFP cells at 37 °C in a humidified incubator with 5% CO_2 for 1 h.

7. Add medium carefully to the top of the Gelfoam®.

8. Incubate 143B-expressing α_v integrin-GFP cells at 37 °C in a humidified incubator with 5% CO_2 for 48–72 h.

9. After 48–72-h seeding, observe cells using an Olympus FV 1000 laser-scanning confocal microscope (Olympus, Tokyo, Japan) with an XLUMPLFL 20× (0.95 NA) water-immersion objective (Olympus, Tokyo, Japan) (*see* **Note 3**) (Fig. 3).

3.5 Fluorescence Imaging

1. Image cultures using an Olympus FW 1000 laser-scanning confocal microscope (Olympus) with a XLUMPLFL 20× (0.95 NA) water-immersion objective.

2. Excite GFP at 488 nm and Z-slices are acquired every 1 um (*see* **Note 4**).

3. Produce images with FV 10-ASW Fluoview software (Olympus) and ImageJ (NIH).

4. Adjust intensity levels of the images.

4   Results

4.1   Cancer Cell Growth in Monolayer Culture

143B cells expressing a_v integrin-GFP have bright fluorescence in the cytoplasm in monolayer culture (Fig. 1). The cells grew as a monolayer without structure.

4.2   Cancer Cell Growth in Matrigel™ Culture

143B a_v integrin-GFP cells were embedded within Matrigel™ (Fig. 2). Colonies of cells were formed within 24 h after seeding. Colonies were seen by 72 h after seeding, but no three-dimensional tissue-like structures were formed.

4.3   Cancer Cell Gelfoam® Histoculture

To investigate the behavior of a_v integrin-GFP on a sponge matrix, cancer cells were seeded on Gelfoam® (Fig. 3). The cancer cells formed three-dimensional tissue-like structures along the Gelfoam® within 72 h (Fig. 3).

4.4   Image Rendering

Images of 143B a_v integrin-GFP cells on each of these substrates were rotated over 85° to visualize tissue-like morphology as well as three-dimensional structures (Fig. 4). The 143B a_v integrin-GFP cells on plastic were observed to behave as individual cells lying flat on the plastic surface of the culture plate. The 143B a_v integrin-GFP cells in Matrigel™ aggregated but did not appear tissue-like. The 143B a_v integrin-GFP cells on Gelfoam® appeared to have a tissue-like structure, which was three-dimensional. The relatively undifferentiated appearance of the tissue-like structure on Gelfoam® resembles 143B tumors formed in vivo [23].

The behavior of 143B osteosarcoma cells in Gelfoam® culture is remarkably different from that of the cells in monolayer culture [24] or in Matrigel™. Tissue-like three-dimensional structures were observed only in Gelfoam® culture. Gelfoam® culture is different than Matrigel™ culture despite the presence of extracellular matrices in Matrigel™.

Despite announcements by Jacks and Weinberg [16] and Abbott [17] about "biology's new dimension" with the use of Matrigel™, sponge-gel matrix histoculture, of which Gelfoam® histoculture is an example, was conceived by Leighton in the early 1950s [15], and allows three-dimensional tissue-like structures to form. This report distinguishes Matrigel™ culture, where only colonies of cells form, from Gelfoam® sponge-gel matrix culture, where three-dimensional tissue-like structures are formed by cancer cells. The cancer cells are readily visualized by a_v integrin GFP expression.

Gelfoam® has also been used to maintain tissue over relatively long periods of time in order to obtain biologically-relevant information concerning tumor biology and drug response (please *see* other chapters of the present volume) [25–27]. Additional uses

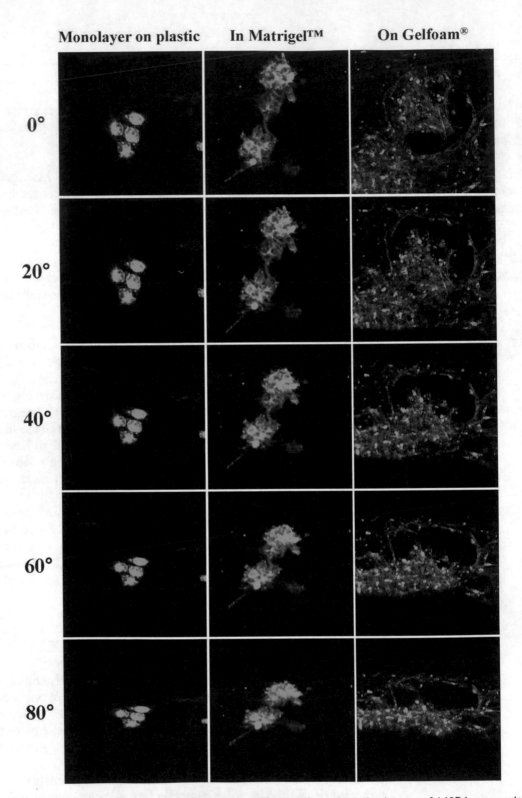

Fig. 4 Image rendering. Phase-contrast fluorescence micrographs and rotating images of 143B human osteosarcoma cells expressing a$_v$ integrin on plastic plates (left panels), in Matrigel™ culture (middle panels), in Gelfoam® culture (right panels) using the Olympus FV1000. Z-stack in 1 mm steps. The depths of images are: on the plastic surface 20 μm; Matrigel™ 70 μm; Gelfoam® 180 μm. These images are rotated on theZ-axis from 0° to 80° in five sections, using the FV-10 ASW software system [22]

of Gelfoam® histoculture include skin and hair growth [28], immune tissue histoculture and HIV infection [29], and nerve growth from stem cells [30, 31] (please *see* Chapters 14, 15, 16, and 17 of the present volume).

The results in this report suggest that further studies be undertaken to determine an optimal general substrate for culture of cancer cells such that they can have as much fidelity as possible to in vivo tumor growth, and that Gelfoam® and similar flexible sponge-matrix substrates are promising candidates to achieve this goal.

Despite the widespread use of Matrigel™ as 3D culture (please *see* the Afterword of this volume), these results as well as Kobayashi et al. [32] demonstratethat Matrigel™ is not true 3D culture.

5   Notes

1. Image the bottom of plastic dish with careful focusing so that α_v integrin foci could be visualized.

2. When observing cancer cells, add RPMI 1640 medium containing 10% FBS on top of the cell culture and image with a water-immersion objective.

3. To image fine quality of z-stuck, reduce the scanning speed.

4. 24 h before seeding, osteosarcoma cells are maintained without G418.

References

1. Leighton J (1960) The propagation of aggregates of cancer cells: implications for therapy and a simple method of study. Cancer Chemother Rep 9:71–72

2. Leighton J, Kalla R, Turner JM Jr, Fennell RH Jr (1960) Pathogenesis of tumor invasion. II. Aggregate replication. Cancer Res 20:575–586

3. Leighton J (1959) Aggregate replication, a factor in the growth of cancer. Science 129(3347):466–467

4. Leighton J, Kalla R, Kline I, Belkin M (1959) Pathogenesis of tumor invasion. I. Interaction between normal tissues and transformed cells in tissue culture. Cancer Res 19:23–27

5. Dawe CJ, Potter M, Leighton J (1958) Progressions of a reticulum-cell sarcoma of the mouse in vivo and in vitro. J Natl Cancer Inst 21:753–781

6. Leighton J (1957) Contributions of tissue culture studies to an understanding of the biology of cancer: a review. Cancer Res 17:929–941

7. Kline I, Leighton J, Belkin M, Orr HC (1957) Some observations on the response of four established human cell strains to hydrocortisone in tissue culture. Cancer Res 17:780–784

8. Leighton J, Kline I, Belkin M, Legallais F, Orr HC (1957) The similarity in histologic appearance of some human cancer and normal cell strains in sponge-matrix tissue culture. Cancer Res 17:359–363

9. Leighton J, Kline I, Belkin M, Orr HC (1957) Effects of a podophyllotoxin derivative on tissue culture systems in which human cancer invades normal tissue. Cancer Res 17:336–344

10. Leighton J, Kline I, Belkin M, Tetenbaum Z (1956) Studies on human cancer using sponge-matrix tissue culture. III. The invasive properties of a carcinoma (strain HeLa) as influenced by temperature variations, by conditioned media, and in contact with rapidly growing chick embryonic tissue. J Natl Cancer Inst 16:1353–1373

11. Leighton J, Kline I, Orr HC (1956) Transformation of normal human fibroblasts into histologically malignant tissue in vitro. Science 123:502

12. Leighton J (1954) The growth patterns of some transplantable animal tumors in sponge matrix tissue culture. J Natl Cancer Inst 15:275–293

13. Leighton J, Kline I (1954) Studies on human cancer using sponge matrix tissue culture. II. Invasion of connective tissue by carcinoma (strain HeLa). Tex Rep Biol Med 12:865–873

14. Leighton J (1954) Studies on human cancer using sponge matrix tissue culture. I. The growth patterns of a malignant melanoma, adenocarcinoma of the parotid gland, papillary adenocarcinoma of the thyroid gland, adenocarcinoma of the pancreas, and epidermoid carcinoma of the uterine cervix (Gey's HeLa strain). Tex Rep Biol Med 12:847–864

15. Leighton JA (1951) sponge matrix method for tissue culture; formation of organized aggregates of cells in vitro. J Natl Cancer Inst 12:545–561

16. Jacks T, Weinberg RA (2002) Taking the study of cancer cell survival to a new dimension. Cell 111:923–925

17. Abbott A (2003) Cell culture: biology's new dimension. Nature 424:870–872

18. Kleinman HK, McGarvey ML, Hassell JR, Star VL, Cannon FB, Laurie GW, Martin GR (1986) Basement membrane complexes with biological activity. Biochemistry 25:312–318

19. Kleinman HK, McGarvey ML, Liotta LA, Robey PG, Tryggvason K, Martin GR (1982) Isolation and characterization of type IV procollagen, laminin, and heparan sulfate proteoglycan from the EHS sarcoma. Biochemistry 21:6188–6193

20. Benton G, Kleinman HK, George J, Arnaoutova I (2011) Multiple uses of basement membrane-like matrix (BME/Matrigel™) in vitro and in vivo with cancer cells. Int J Cancer 128:1751–1757

21. Folkman J, Moscona A (1978) Role of cell shape in growth control. Nature 273:345–349

22. Tome Y, Uehara F, Mii S, Yano S, Zhang L, Sugimoto N, Maehara H, Bouvet M, Tsuchiya H, Kanaya F, Hoffman RM (2014) 3-dimensional tissue is formed from cancer cells in vitro on Gelfoam®, but not on Matrigel™. J Cell Biochem 115:1362–1367

23. Luu HH, Kang Q, Park JK, Si W, Luo Q, Jiang W, Yin H, Montag AG, Simon MA, Peabody TD, Haydon RC, Rinker-Schaeffer CW, He TC (2005) An orthotopic model of human osteosarcoma growth and spontaneous pulmonary metastasis. Clin Exp Metastasis 22:319–329

24. Cukierman E, Pankov R, Stevens DR, Yamada KM (2001) Taking cell matrix adhesions to the third dimension. Science 294:1708–1712

25. Vescio RA, Redfern CH, Nelson TJ, Ugoretz S, Stern PH, Hoffman RM (1987) In vivo-like drug responses of human tumors growing in three-dimensional, gel-supported, primary culture. Proc Natl Acad Sci U S A 84:5029–5033

26. Furukawa T, Kubota T, Hoffman RM (1995) Clinical applications of the histoculture drug response assay. Clin Cancer Res 1:305–311

27. Kubota T, Sasano N, Abe O, Nakao I, Kawamura E, Saito T, Endo M, Kimura K, Demura H, Sasano H, Nagura H, Ogawa N, Hoffman RM (1995) The chemosensitivity study group for the histoculture drug-response assay. Potential of the histoculture drug response assay to contribute to cancer patient survival. Clin Cancer Res 1:1537–1543

28. Li L, Margolis LB, Paus R, Hoffman RM (1992) Hair shaft elongation, follicle growth, and spontaneous regression in long-term, gelatin sponge-supported histoculture of human scalp skin. Proc Natl Acad Sci U S A 89:8764–8768

29. Glushakova S, Baibakov B, Margolis LB, Zimmerberg J (1995) Infection of human tonsil histocultures: a model for HIV pathogenesis. Nat Med 1:1320–1322

30. Mii S, Duong J, Tome Y, Uchugonova A, Liu F, Amoh Y, Saito N, Katsuoka K, Hoffman RM (2013) The role of hair follicle nestin-expressing stem cells during whisker sensory-nerve growth in long-term 3-D culture. J Cell Biochem 114:1674–1684

31. Mii S, Uehara F, Yano S, Tran B, Miwa S, Hiroshima Y, Amoh Y, Katsuoka K, Hoffman RM (2013) Nestin-expressing stem cells promote nerve growth in long-term 3-dimensional Gelfoam®-supported histoculture. PLoS One 8:e67153

32. Kobayashi HI, Man S, Graham C, Kapitain SJ, Teicher BA, Kerbel RS (1993) Acquired multicellular mediated resistance to alkylating agents in cancer. Proc Natl Acad Sci U S A 90:3294–3298

Chapter 20

Imaging the Governing Step of Metastasis in Gelfoam® Histoculture

Robert M. Hoffman and Takashi Chishima

Abstract

Distant organ colonization by cancer cells is the governing step of metastasis. We review in this chapter the modeling and imaging of organ colonization by cancer cells in Gelfoam® histoculture. ANIP 973 lung cancer cells expressing green fluorescent protein (GFP) were injected intravenously into nude mice, whereby they formed brilliantly fluorescing metastatic colonies on the mouse lung. The seeded lung tissue was then excised and incubated in the three-dimensional Gelfoam® histoculture that maintained the critical features of progressive in vivo organ colonization. Tumor progression was continuously visualized by GFP fluorescence of individual cultures over a 52-day period, during which tumor colonies spread throughout the lung. Organ colonization was selective in Gelfoam® histoculture for lung cancer cells to grow on lung tissue, since no growth occurred on histocultured mouse liver tissue. The ability to support selective organ colonization in Gelfoam® histoculture and visualize tumor progression by GFP fluorescence allows the in vitro study of the governing processes of metastasis.

Key words Lung cancer, Gelfoam®, Histoculture, Metastasis, Seed and soil, GFP, Imaging

1 Introduction

Paget, more than 100 years ago, formulated his seed and soil hypothesis that the cells from a given tumor would "seed" only favorable "soil" offered by certain organs [1]. Paget hypothesized that cancer cells must find a suitable "soil" in a target organ (i.e., one that supports colonization) for metastasis to occur. We previously reported that the ability of human colon cancer cells to colonize liver tissue governs whether a particular colon cancer will metastasize to the liver from the colon [2]. This chapter reviews the replication of the seed and soil process in vitro in Gelfoam® histoculture.

Metastatic colony growth by green fluorescent protein (GFP) expressing lung cancer cells in mouse lung was imaged over a 52-day period in the three-dimensional Gelfoam® histoculture. These results provide an invaluable new tool for understanding the

Robert M. Hoffman (ed.), *3D Sponge-Matrix Histoculture: Methods and Protocols*, Methods in Molecular Biology, vol. 1760,
https://doi.org/10.1007/978-1-4939-7745-1_20, © Springer Science+Business Media, LLC, part of Springer Nature 2018

most important steps in tumor host-organ interaction, tumor progression, and metastasis. Because the metastatic colony expansion takes place in vitro, it offers a unique opportunity for developing agents for intervention [3].

2 Materials

1. Fine forceps.

2. Fine scissors.

3. Tissue culture plate.

4. Culture medium (CM): RPMI1640 with L-glutamine and phenol red, 100 µM modified Eagle's medium (MEM) nonessential amino acids (Thermo Fisher Scientific), 1 mM sodium pyruvate, 50 µg/mL gentamycin sulfate, 2.5 µg/mL amphotericin B (Thermo Fisher Scientific), 15% fetal bovine serum (FBS). We recommend testing different batches of FBS before purchasing a large stock.

5. Gelfoam® Absorbable Collagen Sponge USP, 12–7 mm format (Pfizer, NDC: 0009-0315-08) (Gelfoam®) as an example.

6. Gelfoam® trimming aid: double-edged blade.

7. Sterile flat weighing metal spatula.

8. Water-jacketed CO_2 incubator, set at 37 °C, 5% CO_2, ≥90% humidity.

9. Sterile scalpels and blades.

10. Ethanol solution (70%).

11. Methotrexate (MTX) (50 nM).

12. The human lung adenocarcinoma cell line (ANIP 973) (Department of Molecular Biology, Harbin Medical University, Harbin, China) [3].

13. GFP-expression vector containing the codon-optimized hGFP-S65T gene linked to an MTX-resistance gene [4] (CLONTECH, Carlsbad, CA).

14. Nikon microscope equipped with a xenon lamp power supply or a Leica stereo fluorescence microscope model LZ12 equipped with a mercury lamp power supply or equivalent with a GFP filter set(Chroma Technology, Brattleboro, VT).

3 Methods

3.1 Stable Transfection of Lung Cancer Cells with GFP

1. Culture ANIP 973 cells in RPMI 1640 medium (GIBCO) containing 10% fetal calf serum (FCS) (Gemini Biological Products, Calabasas, CA) and 2 mM L-glutamine.

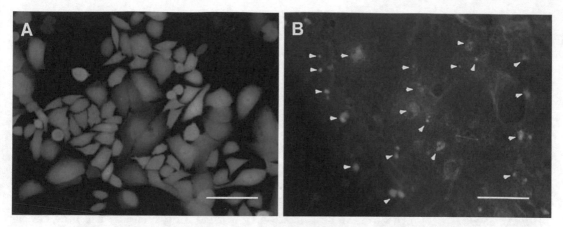

Fig. 1 Mouse lung seeding in vivo by GFP human lung cancer cells. (**a**) GFP-transfected cells selected in 50 nm MTX. Approximately $1.0 \times 3 \times 10^7$ cells were injected into the tail vein of nude mice. (**b**) The excised mouse lung 7 days after cell injection. Numerous fluorescent small colonies can be visualized on the nude mouse lung, with a size of up to approximately 150 microns. Portions of this lung were then explanted into Gelfoam® histoculture (bar = 500 μm) [3]

2. For transfection, incubate near-confluent ANIP 973 cells with a precipitated mixture of LipofectAMINE Reagent (GIBCO) and saturating amounts of the expression vector for 6 h and replenish with fresh medium [5].

3. Harvest ANIP 973 cells with trypsin/EDTA 48 h post-transfection.

4. Subculture the cells at a ratio of 1:15 into selective medium that contained 50 nM methotrexate (MTX).

5. Select cells with a stably integrated expression vector containing the GFP gene by growing transiently transfected cells in the MTX-containing medium.

6. Isolate the clones with cloning cylinders (Bel-Art Products) by trypsin/EDTA.

7. Amplify the clones and transfer by conventional culture methods. Clone-26 was selected due to its high-intensity GFP fluorescence [5] (Fig. 1).

3.2 Tumor Progression in Histoculture

1. Harvest lung tissues aseptically seeded with GFP Clone-26 cells via the tail vein injection from nude mice.

2. Divide lung tissues into pieces of approximately 2–3 mm in diameter and place on prehydrated Gelfoam.

3. Incubate Gelfoam® histoculture at 37 °C in a humidified atmosphere containing 95% air and 5% CO_2.

4. Observe the lung tumor colony growth in the histocultured host lung tissue in the same cultures with fluorescence photomicroscopy of GFP expression at days 6, 14, 24, and 52 of histoculture [3].

5. Use a Nikon microscope equipped with a xenon lamp power supply or a Leica stereo fluorescence microscope model LZ12 equipped with a mercury lamp power supply or equivalent with a GFP filter set[3].

4 Results

4.1 Lung Colony Growth by the GFP-Transfected Lung Tumor Cells in Histoculture

ANIP 973 clone-26-seeded mouse lungs were removed from the mice (*see* **Note 1**) and then histocultured on Gelfoam® (Fig. 1). Tumor colonies grew and spread rapidly in the lung tissue over time in Gelfoam® histoculture. One can visualize continuously the progressive colonization of normal lung tissue by the lung cancer cells in individual cultures by their strong GFP expression (Fig. 2a) (*see* **Note 2**). By day 14, very extensive growth of the colonies had occurred with three different areas of the histocultured lung (Fig. 2b). By day 24 of histoculture, the tumor colonies had grown significantly, reaching sizes of 750 μm in diameter and involving approximately one-half of the histocultured mouse lung (Fig. 2c). By 52 days of histoculture, tumor cells had involved the lung even more extensively and appeared to form multiple layers and histological structures on the histocultured lung (Fig. 2d) (*see* **Note 2**). Also, by day 52, GFP-expressing satellite tumor colonies formed in the Gelfoam distant from the primary colonies in the lung tissue (Fig. 2d) (*see* **Notes 3 and 4**). Figure 2e represents a parallel histoculture at day 24, in which the colonizing lung tumor is brightly visible on the normal mouse lung [3].

5 Notes

1. GFP gene expression described here is bright, heritable, and stable. The cancer cells can be visualized by GFP expression growing on the Gelfoam® histocultured lung for at least 52 days [3].

2. GFP expression in the ANIP 973 lung cancer cells seeded in vivo on the mouse lung was used to visualize the colonization of mouse lung starting from microcolonies and progressing to major colonies. These occupied the majority of the host lung tissues after 52 days in Gelfoam® histoculture [3].

3. The GFP-transfected tumor cells also invaded and colonized the Gelfoam® [5].

4. Gelfoam® histoculture and GFP imaging [6–11] provide the opportunity to readily identify agents which inhibit colonization on the lung by lung cancer cells and since this is the governing step of metastasis, the novel agents have the potential to become anti-metastatic therapeutics.

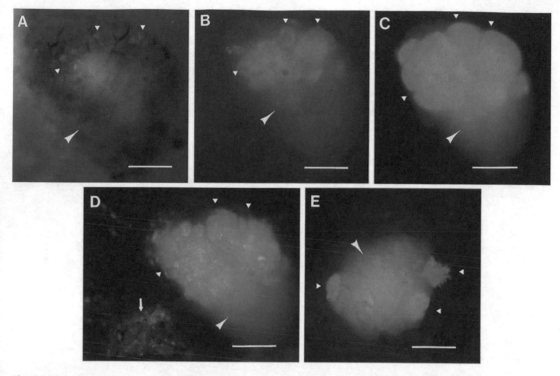

Fig. 2 Mouse lung colonization by GFP human lung tumor cells in Gelfoam® histoculture. (**a**) Growth of the lung cancer cells seeded in the mouse lung after 6 days of histoculture. The tumor colonies still remain as microcolonies with an average size of approximately 90 mm. By day 14, however, the tumor colonies had grown very significantly and occupied a significant fraction of the lung, with the largest individual colony being approximately 400 μm across its largest diameter (**b**). By day 24, very brightly fluorescing tumor colonies had grown to occupy approximately one-half of the histocultured lung, with sizes up to 750 μm (**c**). A parallel, 24-day histocultured lung had brightly fluorescing tumor colonies that invaded the supporting Gelfoam® matrix as well as the lung itself (**e**). (**d**) At 52 days of histoculture, the tumor colonies have grown more on the histocultured mouse lung, with perhaps a secondary layer of colonies forming a structure on top of the first layer of colonies. The colonies were differentiating rather than expanding at this point. Satellite colonies by then had grown into the supporting Gelfoam® matrix (bar = 500 μm) [3]

References

1. Paget S (1889) The distribution of secondary growths in cancer of the breast. Lancet 133(3421):571–573

2. Kuo T, Kubota T, Watanabe M, Furukawa T, Teramoto T, Ishibiki K, Kitajima M, Moossa AR, Penman S, Hoffman RM (1995) Liver colonization competence governs colon cancermetastasis. Proc Natl Acad Sci U S A 92:12085–12089

3. Chishima T, Yang M, Miyagi Y, Li L, Tan Y, Baranov E, Shimada H, Moossa AR, Penman S, Hoffman RM (1997) Governing step of metastasis visualized *in vitro*. Proc Natl Acad Sci U S A 94:11573–11576

4. Zolotukhin S, Potter M, Hauswirth WW, Guy J, Muzycka N (1996) A "humanized" green fluorescent protein cDNA adapted for high-level expression in mammalian cells. J Virol 70:4646–4654

5. Chishima T, Miyagi Y, Wang X, Yang M, Tan Y, Shimada H, Moossa AR, Hoffman RM (1997) Metastatic patterns of lung cancer visualized live and in process by green fluorescence protein expression. Clin Exp Metastasis 15:547–552

6. Hoffman RM (2005) The multiple uses of fluorescent proteins to visualize cancer in vivo. Nature Reviews Cancer 5:796–806

7. Hoffman RM, Yang M (2006) Subcellular imaging in the live mouse. Nature Protocols 1:775–782

8. Hoffman RM, Yang M (2006) Color-coded fluorescence imaging of tumor-host interactions. Nature Protocols 1:928–935

9. Hoffman RM, Yang M (2006) Whole-body imaging with fluorescent proteins. Nature Protocols 1:1429–1438

10. Hoffman RM (2015) Application of GFP imaging in cancer. Laboratory Investigation 95:432–452

11. Hoffman RM, editor (2012) In vivo cellular imaging using fluorescent proteins. Methods and Protocols. Methods in Molecular Biology, Vol. 872, Springer Protocols (Humana Press). ISBN 978-1-61779-796-5

Afterword

Robert M. Hoffman

Organoid Culture: Going Backward In Order To Go Forward

This volume demonstrates the capability of 3D sponge-matrix Gelfoam® histoculture. The volume describes Joe Leighton's pioneering studies of sponge-matrix histoculture in the 1950s [1–15] and subsequent work, spanning almost 70 years. The chapter in this book that shows Gelfoam® histoculture enables the formation and maintenance of tissue structure and function in culture, the ability of cancer and normal cells to self-organize into organoids, and the ability of cancer cells to invade and grow as cell aggregates which replicate themselves as occurs in metastasis in vivo. The volume also describes the ability of stem cells to form nerves, hair, and other structures in Gelfoam® histoculture, and the ability to grow antibody-producing lymphoid tissue that can be infected with HIV in Gelfoam® histoculture.

However, for approximately a half-century from Leighton's initial description of sponge-matrix histoculture, monolayer cell culture was the method of choice for the exploding field of "molecular biology," especially for cancer research [16].

In the middle of the 1980s, Hynda Kleinman and her group [17] further developed a laminin-rich extracellular matrix produced by Engelbreth-Holm-Swarm (EHS) mouse sarcoma cells [18], a substrate for cell culture. The extracellular matrix became commercially known as Matrigel™ and began to be used for what was termed "3-dimensional culture." Matrigel differs from a sponge-matrix such as Gelfoam®, which is gelatinized pig-skin, in that Matrigel is not a sponge but serves as a coating in a cell-culture well or dish. In 1978, Folkman and Moscona [19] demonstrated that shape is critical for the proper "read-out" of gene expression. This may be due to the positions that individual chromosomes take within a cell, depending on their shape [20, 21].

A coating such as Matrigel would not allow cells or a tissue fragment to take their natural shape as would a flexible sponge-matrix such as Gelfoam®. We therefore termed Matrigel 2.5D culture [22]. In Chapter 19 of this volume, where Matrigel and Gelfoam® culture were compared head-to-head using GFP-expressing human 143B osteosarcoma cells, the cells only formed aggregates on Matrigel in contrast to 3-dimensional structures on

Robert M. Hoffman (ed.), *3D Sponge-Matrix Histoculture: Methods and Protocols*, Methods in Molecular Biology, vol. 1760, https://doi.org/10.1007/978-1-4939-7745-1, © Springer Science+Business Media, LLC, part of Springer Nature 2018

Gelfoam®. Nevertheless, Matrigel became the almost completely dominant substrate for "3D culture" beginning in the late 1980s.

Sponge-matrix (Gelfoam) histocultures began to be forgotten except in a few laboratories in the USA (ours) and that of L.B. Margolis (please see Chapter 17) and more in Japan and Korea (please see Chapters 7 and 8 in this volume). To indicate the extent of how sponge-matrix histoculture was forgotten, or never learned, a paper was published in 2002 from Bob Weinberg's laboratory entitled "Taking the Study of Cancer-Cell Survival to a New Dimension" [23], which described two papers from the laboratory of Mina Jihan Bissell [24, 25] which used Matrigel to demonstrate breast cancer cells in what Bissell termed 3D-culture.

Bissell et al. claimed that the breast cancer cells became more "normal" in Matrigel culture after treatment with the anti-integrin antibodies which may be a misinterpretation due to growth arrest of the cancer cells in 2.5D culture.

Bissell stated that "in addition to the cell source, the composition of two basic components must be considered when one established physiologically relevant 3D culture models, the extracellular matrix (ECM) and the medium." No comment was made concerning the need of a substrate that would enable cells to take their normal shape [26].

A news article in Nature [16] followed Weinberg's article and 3D culture was born again! Sponge-matrix histoculture and Joe Leighton were forgotten, or never learned, in the world of current "opinion" leaders who would come to dominate the field. Indeed, the "new" 3D culture paradigm was in reality 2.5D (please see Chapter 19 in this volume). The drug resistance (apoptosis resistance) that Bissell et al. observed in the breast cancer cells in Matrigel culture noted above may be due to most of the malignant and normal cells in the 2.5D culture entering G_0/G_1 phase due to lack of access of nutrients in the center of cell aggregates that were formed. Folkman and Hochberg [27] describe the limit of diffusion in cell aggregates (spheroids) in vitro, thereby limiting cell proliferation only to the outer surface of the mass. We subsequently showed by FUCCI color-coded cell-cycle imaging that the inner cells of such a mass entered G_0/G_1 [28] and thereby become drug resistant (please see Chapter 12 in this volume).

Bissell in a 2017 article [29] entitled "Organoids: A Historical Perspective of Thinking in Three Dimension," claimed to "summarize how work over the past century generated the conceptual framework that has allowed us to make progress in the understanding of tissue-specific morphogenetic programs." However, in her "historical perspective," there is no mention of Leighton's seminal sponge-matrix culture, true 3D culture in which "organoids" were grown almost 70 years ago. Bissell correctly complains that the term "organoid" has now come to imply that it is a completely new field. However, Bissell goes on to state "the first use of the words 'three-dimensional culture models' we believe started with the assays developed by us (Barcells-Hoff et al. [30] and Peterson [31])." This ignorance of the literature is breathtaking, when a simple Google search would uncover numerous such uses of 3D culture models [32–39].

Bissell goes on to list various current definitions of organoids, the most extreme of which is Clevers' [40] which states, "an organoid is now defined as a 3D structure grown from stem cells and consisting of organ-specific cell types that self-organize through cell sorting and spatially-restricted lineage commitment." Clevers' definition is clearly meant to take away the leadership of the "3D culture" field from Bissell and for Clevers be the gatekeeper of the whole field. Clevers also states which stem cells qualify for his organoid definition. This is reminiscent of attempts by Weissman to define what is called a stem cell [41] in order to limit the field and exclude other cell types which could also qualify as stem cells with other definitions.

In another publication, Bissell again attempts to outline the history of the field of 3D culture [42] and states, "it behooves these newcomers who publish under the title of organoids to familiarize themselves with the history of accomplishments of past pioneers...." Bissell states that the field has arrived and scientists now appreciate that "dimensionality" hugely matters. Leighton realized this almost 70 years ago, but Bissell appears to not know this. At least Bissell tried to put the field in historical context. Clevers [40], sets the beginning of his definition of organoid culture with the culture of keratinocytes on an irradiated monolayer of 3 T3 cells by Green's laboratory [43] claiming that "while the term organoid was not used in these pioneering studies, Rheinwald and Green were the first to reconstitute 3D tissue structures from cultured human stem cells." I wonder if Green had any thought of "organoid" or "stem cells" when he performed his experiments. Clevers [40] goes on to state "organoids" can be initiated from the two main types of stem cells, embryonic stem (ES) and induced pluripotent stem cells (iPS) and organ-restricted stem cells as indicated above. Clevers claims [40] that with these types of stem cells, appropriate growth factors can induce the cells to form any specific type of organ-like structure (in "3D"). Clevers goes on to describe different types of organoids from the stem cells he designates. Most of what Clevers [40] describes from his and other laboratories as organoids appears to be cultured on Matrigel (2.5D culture), including brain structures, retina, pituitary, as well as stomach, lung, thyroid, small intestine, liver, and kidney. Clevers also describes [40] organoids derived from adult stem cells, in particular from the stem cell he has worked on, Lgr5, all on Matrigel, including small intestine and colon, stomach, liver, pancreas, prostate, mammary gland, fallopian tube, taste buds, lung, salivary gland, and esophagus [40]. Clevers [44] is now trying to vary the stiffness of matrices claiming the degree of stiffness can influence differentiation. It would be interesting if Clevers would use Gelfoam for his organoid cultures!

In this volume, there are many examples of true organoids even if they may not satisfy Clevers' definition including tonsils on Gelfoam for the study of HIV infection (Chapter 17) and cancer (please see Chapters 3–11, 13, 18, and 19). Sponge-matrix culture of tumors, dates all the way back to Leighton. Gelfoam® histoculture was used for drug testing of individual patient tumors to determine optimal chemotherapy regimens for each patient (please *see* Chapters 7–11).

Clevers has been awarded the $3 Million "Breakthrough Prize" (2013) for his work on organoids [45]. Absence of Leighton's great pioneering work from the current "field of knowledge" is somewhat reminiscent of how patient-derived xenograft models of cancer "disappeared" for almost 20 years and then came back in their 1969 guise about 10 years ago, missing important developments along the way that the current "leaders" of the field never learned [46, 47].

With 3D culture, it seems three-four generations of scientists in the field never learned sponge-matrix histoculture despite numerous papers in Science, PNAS, and so forth. Please see the prophetic quotes from Joseph Leighton in the Foreword of this volume.

Despite the advent of stem cells as they are now called, numerous growth and anti-growth factors, oncogenes, suppressor genes, and so forth, we are still in the era of "Matrigel" culture which as can be seen in Chapter 19 is 2.5D culture. We must go backward to Leighton to move the 3D culture field forward which hopefully will demonstrate the full potential of organoids.

Hopefully, this volume may have an influence to set the course of this field in a proper direction.

References

1. Leighton J (1960) The propagation of aggregates of cancer cells: implications for therapy and a simple method of study. Cancer Chemother Rep 9:71–72

2. Leighton J, Kalla Rl, Turner JM Jr, Fennell RH Jr (1960) Pathogenesis of tumor invasion. II. Aggregate replication. Cancer Res 20:575–586

3. Leighton J (1959) Aggregate replication, a factor in the growth of cancer. Science 129(3347):466–467

4. Leighton J, Kalla Rl, Kline I, Belkin M (1959) Pathogenesis of tumor invasion. I. Interaction between normal tissues and transformed cells in tissue culture. Cancer Res 19(1):23–27

5. Dawe CJ, Potter M, Leighton J (1958) Progressions of a reticulum-cell sarcoma of the mouse in vivo and in vitro. J Natl Cancer Inst 21(4):753–781

6. Leighton J (1957) Contributions of tissue culture studies to an understanding of the biology of cancer: a review. Cancer Res 17(10):929–941

7. Kline I, Leighton J, Belkin M, Orr HC (1957) Some observations on the response of four established human cell strains to hydrocortisone in tissue culture. Cancer Res 17(8):780–784

8. Leighton J, Kline I, Belkin M, Legallais F, Orr HC (1957) The similarity in histologic appearance of some human cancer and normal cell strains in sponge-matrix tissue culture. Cancer Res 17(5):359–363

9. Leighton J, Kline I, Belkin M, Orr HC (1957) Effects of a podophyllotoxin derivative on tissue

culture systems in which human cancer invades normal tissue. Cancer Res 17(4):336–344

10. Leighton J, Kline I, Belkin M, Tetenbaum Z (1956) Studies on human cancer using sponge-matrix tissue culture. III. The invasive properties of a carcinoma (strain HeLa) as influenced by temperature variations, by conditioned media, and in contact with rapidly growing chick embryonic tissue. J Natl Cancer Inst 16(6):1353–1373

11. Leighton J, Kline I, Orr HC (1956) Transformation of normal human fibroblasts into histologically malignant tissue in vitro. Science 123(3195):502

12. Leighton J (1954) The growth patterns of some transplantable animal tumors in sponge matrix tissue culture. J Natl Cancer Inst 15(2):275–293

13. Leighton J, Kline I (1954) Studies on human cancer using sponge matrix tissue culture. II. Invasion of connective tissue by carcinoma (strain HeLa). Tex Rep Biol Med 12(4):865–873

14. Leighton J (1954) Studies on human cancer using sponge matrix tissue culture. I. The growth patterns of a malignant melanoma, adenocarcinoma of the parotid gland, papillary adenocarcinoma of the thyroid gland, adenocarcinoma of the pancreas, and epidermoid carcinoma of the uterine cervix (Gey's HeLa strain). Tex Rep Biol Med 12(4):847–864

15. Leighton J (1951) A sponge matrix method for tissue culture; formation of organized aggregates of cells in vitro. J Natl Cancer Inst 12(3):545–561

16. Abbott A (2003) Cell culture: Biology's new dimension. Nature 424:870–872

17. Kleinman HK, McGarvey ML, Hassell JR, Star VL, Cannon FB, Laurie GW, Martin GR (1986) Basement membrane complexes with biological activity. Biochemistry 25(2):312–318

18. Orkin RW, Gehron P, McGoodwin EB, Martin GR, Valentine T, Swarm R (1977) A murine tumor producing a matrix of basement membrane. J Exp Med 145(1):204–220

19. Folkman J, Moscona A (1978) Role of cell shape in growth control. Nature 273(5661):345–349

20. Wang Y, Nagarajan M, Uhler C, Shivashankar GV (2017) Orientation and repositioning of chromosomes correlate with cell geometry-dependent gene expression. Mol Biol Cell 28(14):1997–2009

21. Sehgal N, Fritz AJ, Vecerova J, Ding H, Chen Z, Stojkovic B, Bhattacharya S, Xu J, Berezney R (2016) Large-scale probabilistic 3D organization of human chromosome territories. Hum Mol Genet 25(3):419–436

22. Tome Y, Uehara F, Mii S, Yano S, Zhang L, Sugimoto N, Maehara H, Bouvet M, Tsuchiya H, Kanaya F, Hoffman RM (2014) 3-dimensional tissue is formed from cancer cells in vitro on Gelfoam®, but not on Matrigel™. J Cell Biochem 115:1362–1367

23. Jacks T, Weinberg RA (2002) Taking the study of cancer cell survival to a new dimension. Cell 111(7):923–925

24. Weaver VM, Lelièvre S, Lakins JN, Chrenek MA, Jones JC, Giancotti F, Werb Z, Bissell MJ (2002) beta4 integrin-dependent formation of polarized three-dimensional architecture confers resistance to apoptosis in normal and malignant mammary epithelium. Cancer Cell 2(3):205–216

25. Weaver VM, Petersen OW, Wang F, Larabell CA, Briand P, Damsky C, Bissell MJ (1997) Reversion of the malignant phenotype of human breast cells in three-dimensional culture and in vivo by integrin blocking antibodies. J Cell Biol 137(1):231–245

26. Schmeichel KL, Bissell MJ (2003) Modeling tissue-specific signaling and organ function in three dimensions. J Cell Sci 116(Pt 12):2377–2388

27. Folkman J, Hochberg M (1973) Self-regulation of growth in three dimensions. J Exp Med 138(4):745–753

28. Yano S, Tazawa H, Hashimoto Y, Shirakawa Y, Kuroda S, Nishizaki M, Kishimoto H, Uno F, Nagasaka T, Urata Y, Kagawa S, Hoffman RM, Fujiwara T (2013) A genetically engineered oncolytic adenovirus decoys and lethally traps quiescent cancer stem-like cells into $S/G_2/M$ phases. Clin Cancer Res 19:6495–6505

29. Simian M, Bissell MJ (2017) Organoids: a historical perspective of thinking in three dimensions. J Cell Biol 216(1):31–40

30. Barcells-Hoff MH, Aggeler J, Ram TG, Bissell MJ (1989) Functional differentiation and alveolar morphogenesis of primary mammary cultures on reconstituted basement membrane. Development 105:223–235

31. Petersen OW, Ronnov-Jessen L, Howlett AR, Bissell MJ (1992) Interaction with basement membrane serves to rapidly distinguish growth and differentiation pattern of normal and malignant human breast epithelial cells. Proc Natl Acad Sci U S A 89:9064–9068

32. Li L, Hoffman RM (1991) Eye tissues grown in 3-dimensional histoculture for toxicological studies. J Cell Pharmacol 2:311–316

33. Guadagni F, Li L, Hoffman RM (1992) Targeting antibodies to live tumor tissue in 3-D histoculture. In Vitro Cell Dev Biol 28A:297–299

34. Mii S, Duong J, Tome Y, Uchugonova A, Liu F, Amoh Y, Saito N, Katsuoka K, Hoffman RM (2013) The role of hair follicle nestin-expressing stem cells during whisker sensory-nerve growth in long-term 3D culture. J Cell Biochem 114:1674–1684

35. Yano S, Miwa S, Mii S, Hiroshima Y, Uehara F, Yamamoto M, Kishimoto H, Tazawa H, Bouvet M, Fujiwara T, Hoffman RM (2014) Invading cancer cells are predominantly in G_0/G_1 resulting in chemoresistance demonstrated by real-time FUCCI imaging. Cell Cycle 13:953–960

36. Tome Y, Uehara F, Mii S, Yano S, Zhang L, Sugimoto N, Maehara H, Bouvet M, Tsuchiya H, Kanaya F, Hoffman RM (2014) 3-dimensional tissue is formed from cancer cells in vitro on Gelfoam® but not on Matrigel™. J Cell Biochem 115:1362–1367

37. Zhang L, Wu C, Bouvet M, Yano S, Hoffman RM (2015) Traditional Chinese medicine herbal mixture LQ arrests FUCCI-expressing HeLa cells in G0/G1 phase in 2D plastic, 2.5-D Matrigel®, and 3D Gelfoam® culture visualized with FUCCI imaging. Oncotarget 6:5292–5298

38. Yano S, Miwa S, Mii S, Hiroshima Y, Uehara F, Kishimoto H, Tazawa H, Zhao M, Bouvet M, Fujiwara T, Hoffman RM (2015) Cancer cells mimic in vivo spatial-temporal cell-cycle phase distribution and chemosensitivity in 3-dimensional Gelfoam® histoculture but not 2-dimensional culture as visualized with real-time FUCCI imaging. Cell Cycle 14:808–819

39. Yano S, Takehara K, Miwa S, Kishimoto H, Tazawa H, Urata Y, Kagawa S, Bouvet M, Fujiwara T, Hoffman RM (2017) GFP labeling kinetics of triple-negative human breast cancer by a killer-reporter adenovirus in 3D Gelfoam® histoculture. In Vitro Cell Dev Biol Anim 53:479–482

40. Clevers H (2016) Modeling development and disease with organoids. Cell 165:1586–1597

41. Wagers AJ, Weissman IL (2004) Plasticity of adult stem cells. Cell 116:639–648

42. Bissell MJ (2017) Goodbye flat biology—time for the 3rd and the 4th dimensions. J Cell Sci 130:3–5

43. Rheinwald JG, Green H (1975) Formation of a keratinizing epithelium in culture by a cloned cell line derived from a teratoma. Cell 6:317–330

44. Gjorevski N, Sachs N, Manfrin A, Giger S, Bragina ME, Ordóñez-Morán P, Clevers H, Lutolf MP (2016) Designer matrices for intestinal stem cell and organoid culture. Nature 539:560–564

45. https://en.wikipedia.org/wiki/Breakthrough_Prize_in_Life_Sciences

46. Hoffman RM (2015) Patient-derived orthotopic xenografts: better mimic of metastasis than subcutaneous xenografts. Nat Rev Cancer 15:451–452

47. Hoffman RM (ed) (2017) Patient-derived mouse models of cancer. Molecular and translational medicine (Coleman WB, Tsongalis GJ, Series eds). ISSN: 2197–7852

Epilogue

Robert M. Hoffman

Whither Gelfoam® Histoculture

Science does not usually move forward in a straight line. As we have seen in this volume, sometimes science has to go backward in order to move forward.

It is the Editor's hope that this volume will serve to inform workers in cell culture of the great advantage of sponge-gel histoculture and that many will use it to advance their field.

As we have seen in this volume, sponge-gel histoculture has many applications:

1. We can read in Chapter 15 that Gelfoam® histoculture supports the growth of hair follicle-associated pluripotent (HAP) stem cells that can be used for spinal cord repair in mice. This may be a possible clinical approach for spinal cord repair.

2. We can read in Chapter 16 that HAP stem cells form growing nerves in Gelfoam® histoculture that can interact with other nerves in the same culture. It is possible that these results will be used clinically to grow nerves for regenerative medicine and possibly to develop an in vitro nervous system.

3. We can read in Chapter 14 that Gelfoam® histoculture can support hair growth from intact skin, including scalp or from isolated follicles that have very active HAP stem cells. It is possible that novel agents or procedures will be discovered in Gelfoam® histoculture to stimulate hair growth. This will result in improved quality of life for millions.

4. We can read in Chapter 17 that lymphoid tissue can function in Gelfoam® histoculture producing specific antibodies in response to antigens and that HIV can infect histocultured lymphoid tissues. It is possible that new means to stimulate production of specific antibodies will be discovered in Gelfoam® histoculture and also that we may achieve an understanding the true nature of AIDS by the study of lymphoid tissue in Gelfoam® histoculture.

5. We can read in Chapter 20 that organ colonization by cancer cells can be replicated in Gelfoam® histoculture. Distant organ colonization is the governing step of metastasis and that we can further

Robert M. Hoffman (ed.), *3D Sponge-Matrix Histoculture: Methods and Protocols*, Methods in Molecular Biology, vol. 1760, https://doi.org/10.1007/978-1-4939-7745-1, © Springer Science+Business Media, LLC, part of Springer Nature 2018

understand its mechanism and prevent or reverse this step in Gelfoam® histoculture and apply this knowledge in the clinic.

6. In Chapter 12, we can read about imaging the phase of the cell cycle of each cancer cell in Gelfoam® histoculture using "FUCCI" imaging. It is possible that Gelfoam® histoculture will enable further understanding of the aberrations of the cancer cell cycle, leading to more effective cancer therapy.

7. We can read in Chapter 13 that the general metabolic defect of methionine dependence in cancer cells is expressed in Gelfoam® histoculture. This should enable the further understanding of methionine dependence and its fundamental relationship to the very nature of cancer which may lead to a general cure.

8. In Chapter 18, we can read that DNA repair can be imaged after UV light irradiation of cancer cells in Gelfoam® histoculture. It is possible that a deeper understanding of the relationship of DNA repair and cancer and its therapy will be found in Gelfoam® histoculture.

9. In Chapter 3, we can read that tumors in Gelfoam® histoculture grow similar to how they grew in the patients, including forming aberrant glandular and other structures as well as their invasive ability. The ability to grow tumors in vitro such that they replicate their in vivo growth and invasive properties should allow a deeper understanding of the nature of cancer and the possibility of new drug discovery for metastasis.

10. In Chapter 4, we can read that tumor antigen expression in Gelfoam® histoculture mimics antigen expression in vivo. We can also read in Chapter 4 that tumor antigens in Gelfoam® histoculture can be targeted with antibodies. This method can be used to identify new antibody cancer therapeutics such as those that can be used in immuno-oncology.

11. In Chapter 6, we can read how tumors in Gelfoam® histoculture can be used for cancer diagnosis by their cell proliferation index. This technology can lead to "real-time" cancer pathology diagnosis of many, in not all, cancer types as a new paradigm improving upon the "dead pathology" era we have been in for more than a century since the time of Virchow.

12. In Chapters 7–10, we can read about clinical correlation of drug response in the Gelfoam®-supported histoculture drug response assay (HDRA) to outcome of patients with gastrointestinal cancer, ovarian cancer, head and neck cancer, and breast cancer. Future experiments will involve correlation, retrospectively or prospectively, of HDRA results with clinical outcome in other cancers.

Application of the histoculture drug response assay for lung cancer has been preliminarily investigated. Chemosensitivities of lung cancer HDRA tissues to cisplatinum (CDDP), doxorubicin (DOX), mitomycin C (MMC), 5-fluorouracil (5-FU), docetaxel (DOC), paclitaxel (PTX), etoposide, irinotecan (IRN), and gemcitabine (GEM) were investigated. Cutoff inhibition rates were determined with each drug for non-small cell lung cancer and were used to calculate predictabilities for chemotherapy responses in the clinic. The evaluability rate of the histoculture drug response assay was high at 97.4%. Predictability includes true-positive and true-negative rates of 73.2% and 100%, respectively, with an accuracy of 83.0%. The histoculture drug response assay may be applicable to non-small cell lung cancer for the prediction of responses to chemotherapy [1].

Future experiments will involve prospective, randomized clinical trials of the HDRA for particular cancers with the goal of making the HDRA "standard of care."

13. In Chapter 11, we can read about the Gelfoam® histoculture of benign prostate hypertrophy (BPH) and prostate cancer and their response to androgen. Future experiments will use this Gelfoam® histoculture system to investigate new androgen blockers of BPH and prostate cancer.

14. Conditions of Gelfoam® histoculture, such as pH of the medium [2], can be modified to determine effects on drug sensitivity. Such medium modifications may be translatable to the clinic.

Gelfoam® histoculture has great potential for cancer, stem cells, immunology, hair growth, and other areas of medicine. We owe much to the pioneer Joe Leighton. We are now approaching 70 years of sponge-gel histoculture. It is hoped the next 70 years will see much progress in medicine and science with sponge-gel histoculture.

References

1. Yoshimasu T, Oura S, Hirai I, Tamaki T, Kokawa Y, Hata K, Ohta F, Nakamura R, Kawago M, Tanino H, Okamura Y, Furukawa T (2007) Data acquisition for the histoculture drug response assay in lung cancer. J Thorac Cardiovasc Surg 133:303–308

2. Flowers JL, Hoffman RM, Driscoll TA, Wall ME, Wani MC, Manikumar G, Friedman HS, Dewhirst M, Colvin OM, Adams DJ (2003) The activity of camptothecin analogues is enhanced in histocultures of human tumors and human tumor xenografts by modulation of extracellular pH. Cancer Chemother Pharmacol 52:253–261

INDEX

Robert M. Hoffman (ed.), *3D Sponge-Matrix Histoculture: Methods and Protocols*, Methods in Molecular Biology, vol. 1760, https://doi.org/10.1007/978-1-4939-7745-1, © Springer Science+Business Media, LLC, part of Springer Nature 2018

Printed in the United States
By Bookmasters